Development of Unconventional Hydrocarbon Resources in the Appalachian Basin: Workshop Summary

Anne Linn, *Rapporteur*

Committee on the Development of Unconventional Hydrocarbon Resources in the Appalachian Basin

Committee on Earth Resources

Board on Earth Sciences and Resources

Water Science and Technology Board

Division on Earth and Life Studies

NATIONAL RESEARCH COUNCIL
OF THE NATIONAL ACADEMIES

THE NATIONAL ACADEMIES PRESS
Washington, D.C.
www.nap.edu

THE NATIONAL ACADEMIES PRESS 500 Fifth Street, NW Washington, DC 20001

NOTICE: The project that is the subject of this report was approved by the Governing Board of the National Research Council, whose members are drawn from the councils of the National Academy of Sciences, the National Academy of Engineering, and the Institute of Medicine. The members of the committee responsible for the report were chosen for their special competences and with regard for appropriate balance.

This study was supported by West Virginia University Research Corporation. The opinions, findings, and conclusions or recommendations contained in this document are those of the author and do not necessarily reflect the views of the organization that provided support for the project. Mention of trade names or commercial products does not constitute their endorsement by the sponsor.

International Standard Book Number-13: 978-0-309-29881-0
International Standard Book Number-10: 0-309-29881-4

Cover: Geologic map of West Virginia (background). SOURCE: Map 25A, West Virginia Geological and Economic Survey. Superimposed are photographs of the following: rectangular joints in the Utica Shale near Fort Plain, New York, copyright by Michael C. Rygel (left); Marcellus Shale outcrop in Highland County, Virginia, photo by James Coleman (center); Marcellus Shale just south of Marcellus, New York (right). SOURCES: Wikipedia Commons (left and right) and the U.S. Geological Society (center).

Additional copies of this report are available for sale from the National Academies Press, 500 Fifth Street, NW, Keck 360, Washington, DC 20001; (800) 624-6242 or (202) 334-3313; http://www.nap.edu/.

Printed in the United States of America.

THE NATIONAL ACADEMIES

Advisers to the Nation on Science, Engineering, and Medicine

The **National Academy of Sciences** is a private, nonprofit, self-perpetuating society of distinguished scholars engaged in scientific and engineering research, dedicated to the furtherance of science and technology and to their use for the general welfare. Upon the authority of the charter granted to it by the Congress in 1863, the Academy has a mandate that requires it to advise the federal government on scientific and technical matters. Dr. Ralph J. Cicerone is president of the National Academy of Sciences.

The **National Academy of Engineering** was established in 1964, under the charter of the National Academy of Sciences, as a parallel organization of outstanding engineers. It is autonomous in its administration and in the selection of its members, sharing with the National Academy of Sciences the responsibility for advising the federal government. The National Academy of Engineering also sponsors engineering programs aimed at meeting national needs, encourages education and research, and recognizes the superior achievements of engineers. Dr. C. D. Mote, Jr., is president of the National Academy of Engineering.

The **Institute of Medicine** was established in 1970 by the National Academy of Sciences to secure the services of eminent members of appropriate professions in the examination of policy matters pertaining to the health of the public. The Institute acts under the responsibility given to the National Academy of Sciences by its congressional charter to be an adviser to the federal government and, upon its own initiative, to identify issues of medical care, research, and education. Dr. Harvey V. Fineberg is president of the Institute of Medicine.

The **National Research Council** was organized by the National Academy of Sciences in 1916 to associate the broad community of science and technology with the Academy's purposes of furthering knowledge and advising the federal government. Functioning in accordance with general policies determined by the Academy, the Council has become the principal operating agency of both the National Academy of Sciences and the National Academy of Engineering in providing services to the government, the public, and the scientific and engineering communities. The Council is administered jointly by both Academies and the Institute of Medicine. Dr. Ralph J. Cicerone and Dr. C. D. Mote, Jr., are chair and vice chair, respectively, of the National Research Council.

www.national-academies.org

COMMITTEE ON THE DEVELOPMENT OF UNCONVENTIONAL HYDROCARBON RESOURCES IN THE APPALACHIAN BASIN

GEORGE M. HORNBERGER, *Chair*, Vanderbilt University, Nashville, Tennessee
KATE HADLEY BAKER, BP (retired), Houston, Texas
SUSAN L. BRANTLEY, Pennsylvania State University, University Park
MICHAEL E. HOHN, State of West Virginia, Morgantown
CARL KIRBY, Bucknell University, Lewisburg, Pennsylvania

National Research Council Staff

ANNE M. LINN, Study Director, Board on Earth Sciences and Resources
LAURA HELSABECK, Senior Program Officer, Water Science and Technology Board (until June 7, 2013)
ERIC J. EDKIN, Senior Program Assistant, Board on Earth Sciences and Resources

COMMITTEE ON EARTH RESOURCES

CLAYTON R. NICHOLS, *Chair*, Idaho National Engineering and Environmental Laboratory (retired), Sandpoint
JAMES A. BRIERLEY, Brierley Consultancy, LLC, Highlands Ranch, Colorado
MICHAEL J. CARROLL, Hunt Oil, Dallas, Texas
THURE CERLING, University of Utah, Salt Lake City
ELAINE T. CULLEN, NIOSH Spokane Research Laboratory (retired), Chattaroy, Washington
DONALD JUCKETT, American Association for Petroleum Geologists (retired), Springfield, Virginia
DOROTHY J. MERRITTS, Franklin & Marshall College, Lancaster, Pennsylvania
MARY M. POULTON, University of Arizona, Tucson
ARTHUR W. RAY, City of Rockville, Maryland
RICHARD J. SWEIGARD, University of Kentucky, Lexington
ELIZABETH J. WILSON, University of Minnesota, Minneapolis

National Research Council Staff

ELIZABETH A. EIDE, Director, Board on Earth Sciences and Resources
NICHOLAS D. ROGERS, Financial and Research Associate, Board on Earth Sciences and Resources
ERIC J. EDKIN, Senior Program Assistant, Board on Earth Sciences and Resources

BOARD ON EARTH SCIENCES AND RESOURCES

National Research Council Staff

Acknowledgments

This workshop summary has been reviewed in draft form by persons chosen for their diverse perspectives and technical expertise in accordance with procedures approved by the National Research Council's Report Review Committee. The purposes of this review are to provide candid and critical comments that will assist the institution in making the published summary as sound as possible and to ensure that the summary meets institutional standards of objectivity, evidence, and responsiveness to the study charge. The review comments and draft manuscript remain confidential to protect the integrity of the deliberative process. We wish to thank the following for their participation in the review of this summary:

James Coleman, U.S. Geological Survey, Reston, Virginia
Sally Entrekin, University of Central Arkansas, Conway
George Hornberger, Vanderbilt University, Nashville, Tennessee

Although the reviewers listed above have provided many constructive comments and suggestions, they were not asked to endorse, nor did they see the final draft, of the workshop summary before its release. The review of this summary was overseen by David Dzombak, Carnegie Mellon University. Appointed by the National Research Council, he was responsible for making certain that an independent examination of this summary was carried out in accordance with institutional procedures and that all review comments were carefully considered. Responsibility for the final content of this summary rests entirely with the author and the National Research Council.

The planning committee would like to thank West Virginia University—particularly James P. Clements, Jay Cole, and Lisa Martin—for sponsoring this activity, hosting the workshop, and providing superb logistical support. Thanks also go to Ray Boswell, Rosemary Capo, Joseph Frantz, Jr., Cliff Frohlich, Jose Fuentes, Paulina Jaramillo, Kelly Maloney, Michael Powelson, John Veil, and Hannah Wiseman, whose excellent plenary presentations provided a strong foundation for discussion. The committee also extends its thanks to the participants who took on multiple roles in the working groups, giving flash talks to start discussion or serving as rapporteurs: Brian Anderson, Richard Bajura, Daniel Billman, Margaret Brittingham, Robert Burruss, Tim Carr, Martin Chapman,

Nigel Clark, Jeffrey Daniels, Patrick Drohan, Mark Engle, Barbara Evans Fleischauer, Gregory Frost, Paulina Jaramillo, Donald Juckett, William Kappel, Zuleima Karpyn, Abby Kinchy, Peter MacKenzie, Kris Nygaard, Gary Slagel, James Van Nostrand, Radisav Vidic, Richard Wilkin, Hannah Wiseman, Gregory Wrightstone, and Paul Ziemkiewicz. Finally, the committee extends its thanks to the honorable John D. Rockefeller IV, U.S. Senator for West Virginia, who made remarks in the form of a letter read at the workshop (Appendix A).

Contents

1

Introduction

Most of the oil and gas produced in the United States comes from conventional reservoirs in which hydrocarbons have accumulated in discrete structural or stratigraphic traps below relatively impermeable rock and above a well-defined hydrocarbon–water interface (USGS, 2002).[1] However, a growing fraction comes from unconventional reservoirs—geographically extensive accumulations of hydrocarbons held in low-permeability rock (e.g., shale, siltstone) with diffuse boundaries and no obvious traps or hydrocarbon–water contacts. Unconventional hydrocarbons include shale gas, tight oil, tight gas, coalbed methane, and gas hydrates.

Shale gas is the fastest growing source of U.S. natural gas (Figure 1.1). The Energy Information Administration projects that shale gas will account for nearly half of U.S. natural gas production in 2040, compared with less than 10 percent in 2011 (EIA, 2013a). However, it is difficult to extract economically. The low-permeability rock holding shale gas and other unconventional hydrocarbon resources is generally hydraulically fractured to free the gas.

Hydraulic fracturing uses a high-pressure injection of fluid (generally water), proppant (often sand), and small amounts of chemicals to create fracture networks or enhance existing fractures in the rocks to stimulate production (e.g., NRC, 2013).[2] The consequences of this practice and other aspects of unconventional hydrocarbon production have been a matter of intense public debate. Proponents of unconventional hydrocarbon development emphasize issues such as greater energy security, economic development, job creation, reduced greenhouse gas emissions from natural gas relative to other fossil fuels, and well-established engineering techniques. Opponents of unconventional hydrocarbon development identify potential problems such as contamination of surface water and groundwater, depletion of water resources, fragmentation and loss of habitat, public health effects, induced seismicity, air pollution, and increased greenhouse gas emissions due to leakage of natural gas. Each state with potential shale gas resources has largely sought its own balance in developing the resource and safeguarding the environment (e.g., Wiseman, 2012). In the Appala-

[1] See also http://energy.cr.usgs.gov/oilgas/addoilgas/.
[2] See also http://energy.usgs.gov/OilGas/UnconventionalOilGas/HydraulicFracturing.aspx.

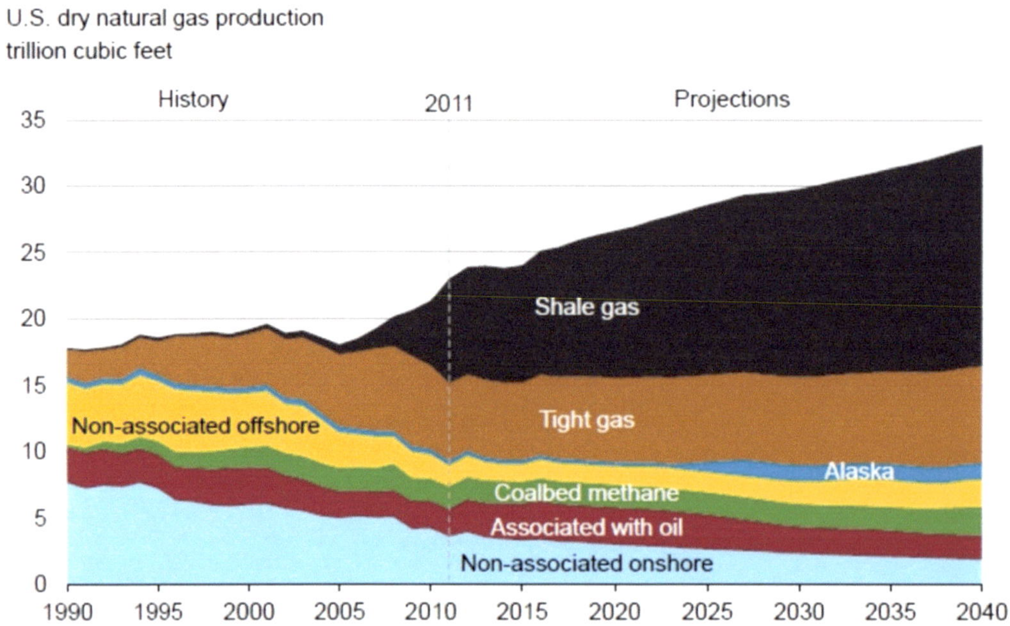

FIGURE 1.1 Trends in U.S. production of dry natural gas by source, in trillion cubic feet, 1990–2011 and projections to 2040. Projections are based on energy factors (production, imports, consumption) and economic factors (prices, economic indicators such as gross domestic product, energy intensity) and assume that current laws and regulations affecting the energy sector remain unchanged through the projection period. SOURCE: EIA (2013a).

chian Basin, for example, shale gas development is proceeding in Ohio, Pennsylvania, and West Virginia, while New York and Maryland have commissioned studies to assess potential impacts.[3]

At the request of West Virginia University, the National Research Council organized a workshop to examine the geology and unconventional hydrocarbon resources of the Appalachian Basin; technical methods for producing unconventional hydrocarbons and disposing of wastewater; the potential effects of production on the environment; relevant policies and regulations; and priorities for future scientific and engineering research (see Box 1.1). A comprehensive treatment of these topics is not possible in a 2-day workshop, so the planning committee chose to focus on shale gas and tight gas, which are economically important to the region and also of intense public interest. West Virginia University expects to use the results of the workshop to support its land-grant university mission of providing new knowledge, reaching out to the community, and creating economic development opportunities.[4]

The Appalachian Basin includes several major shale gas and tight gas formations at different depths and spatial extents (e.g., Coleman et al., 2011; Kirschbaum et al., 2012). The boundaries of the most productive formations are shown in Figure 1.2. Some of the shale gas and tight gas formations produce methane (dry gas), and others produce methane mixed with natural gas liquids such as ethane, propane, and butane (wet gas). These natural gas liquids can be separated from the methane and sold as separate products for a variety of industrial applications.

[3] See http://www.dec.ny.gov/regulations/77353.html; http://www.mde.state.md.us/programs/Land/mining/marcellus/Pages/index.aspx.

[4] Presentation by James P. Clements, president of West Virginia University, on September 9, 2013.

BOX 1.1
Statement of Task

An ad hoc NRC committee will organize a 2-day public workshop on development of unconventional hydrocarbon resources in the Appalachian Basin. The workshop will include presentations that examine the numerous geoscientific aspects of hydrocarbon development from unconventional resources, including natural gas, oil, and natural gas liquids. The workshop will provide an independent forum for dialogue, including contributions from experts representing a full range of geoscience and engineering fields. The committee will develop the agenda, select and invite speakers and discussants, and moderate the discussions.

Topics of emphasis in the workshop will include:

1. Geology and hydrocarbon resources—Main hydrocarbon-bearing geologic formations in and around the Appalachian Basin, including the Marcellus, Utica, and Devonian shales, and their estimated resources of natural gas, oil, and/or natural gas liquids and current and projected production levels.
2. Potential effects on surface water and groundwater quality and quantity—Connections between hydraulic fracturing and other production technologies and processes, and water systems, including scientific data and methods in assessing impacts.
3. Potential effects on landscapes, including soil and living organisms, and other environmental systems—Connections between hydraulic fracturing and other production technologies and processes on environmental systems, including scientific data and methods in assessing impacts.
4. Technical and engineering processes—Current and prospective technical and engineering processes for exploration and production of hydrocarbons from unconventional resources, and management methods for wastewater, including disposal.

As appropriate, the workshop will also include presentations on relevant state and federal water quality laws, regulations, and permitting processes, as well as relevant land-use and land management policies. Following the workshop, the National Research Council will issue an individually-authored summary of the workshop, prepared by a designated rapporteur. This report will summarize the discussions at the workshop, including priorities for future scientific and engineering research as identified by workshop participants.

OVERVIEW OF THE WORKSHOP

The workshop was organized and convened by a planning committee and held on September 9-10, 2013, at West Virginia University. Participants were drawn from universities, private companies, federal and state government bodies, and nongovernmental organizations to bring a wide range of expertise and perspectives to the workshop. Sixty-six people attended the workshop, and an additional 54 people participated remotely via webcast (Appendix C).[5]

The workshop was organized roughly around the statement of task (Box 1.1), with sessions on the geology, resources, and production in the Appalachian Basin (Task 1); water and regulations (Task 2); and ecosystems, air, and climate (environmental systems, Task 3). Technical and engineering processes (Task 4) were discussed in all three sessions. Each session began with a few plenary talks to provide a broad overview of the topics (see the agenda in Appendix B). Next, participants broke into four multidisciplinary working groups to discuss the issues in more depth. Each set of working groups discussed the same topics. To begin discussion, two working group members gave

[5] The webcast is available at http://www.tvworldwide.com/events/nas/130909/.

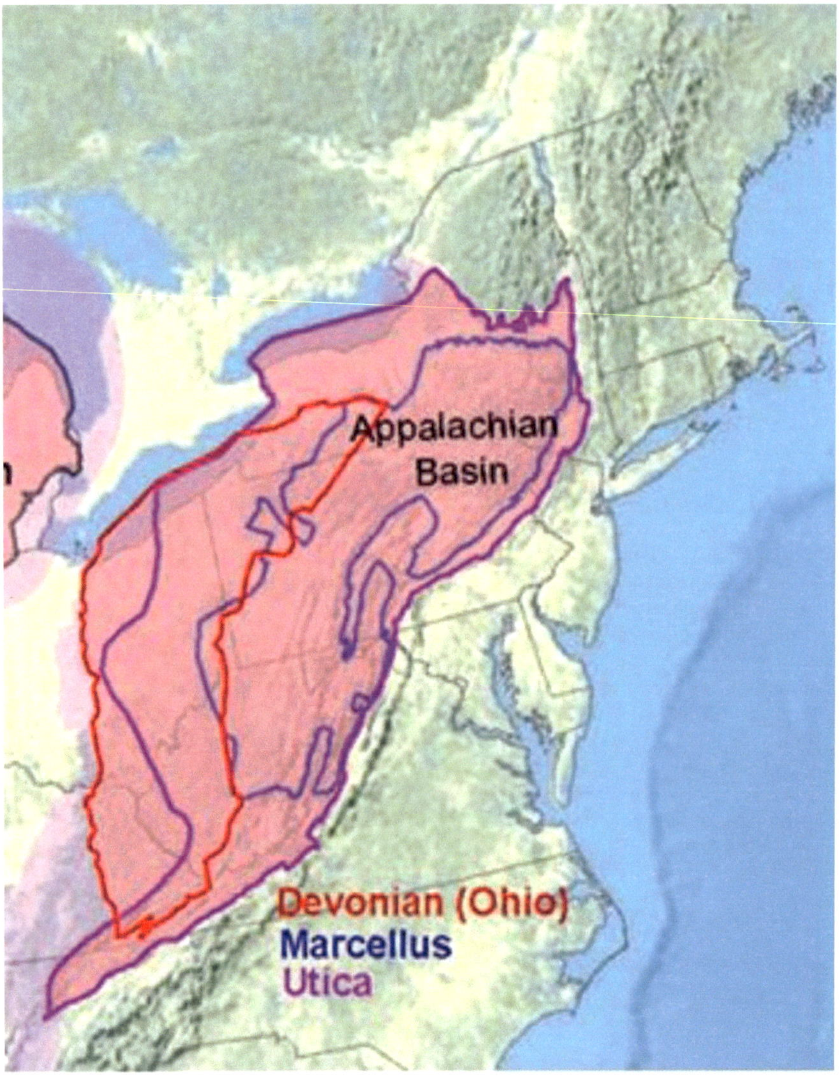

FIGURE 1.2 Extent of three major shale gas formations in the Appalachian Basin: the Marcellus, Utica, and Ohio shales. SOURCE: Energy Information Administration.

brief talks, one from an industry perspective and one from a nonindustry perspective, commenting on the plenary presentations and raising other important issues. The key points of discussion, technical and engineering issues, and future research priorities were captured by a rapporteur and presented back in plenary session (see Appendix D). The workshop concluded with some brief thoughts by planning committee members and other workshop participants.

SHALE GAS PRODUCTION PROCESS

The following overview of the shale gas production process was based on sources cited at the workshop[6] or provided as background by the planning committee (e.g., NETL, 2009). The shale gas production process has several stages, most of which are governed by regulations and subjected to various tests by the operator. First, the site is prepared by clearing and leveling the land surface and constructing the production infrastructure, including a well pad for the drilling rig and other equipment, an access road to the well pad, reserve pits to manage drilling fluid and cuttings, and compression stations to facilitate the transport of gas. In Pennsylvania, the average size of a Marcellus Shale well pad is about 3 acres, and an additional 6 acres are occupied by roads, pipelines, and water impoundments (Johnson, 2010). Most pads installed in the past few years have two or three wells, although pads can accommodate up to 12 wells, depending on issues such as lease size and restrictions, available capital, economics, terrain, and logistical challenges (Kuntz et al., 2011; Ladlee and Jacquet, 2011).

Each well is drilled in intervals. The first section of the well is drilled with a large-diameter drill bit, and a section of pipe is inserted into the hole. Cement is pumped into the space between the hole and the pipe to secure the pipe in place. A smaller-diameter hole is then drilled to a depth below the water table. A length of pipe, called the surface casing, is set into the borehole and cemented in place. A deeper interval is then drilled and another casing string is cemented. This step may be repeated several times with additional concentric strings of casing (Figure 1.3). Below the aquifers, the casing is set to ensure that gas from the producing zone flows into the well and not into other low-pressure zones outside of the casing.

At depths slightly above the shale layer, the wellbore is generally turned, and drilling continues horizontally through the shale layer for several thousand feet. Horizontal drilling significantly increases the well's exposure to the gas-producing formation, thereby increasing production. Next, perforating guns are lowered into the producing section of the well. Explosive charges are detonated to puncture holes through the cement, casing, and edge of the rock formation. This is followed by hydraulic fracturing. Rather than perforating and fracturing the entire gas-producing interval at one time, the process is generally performed on smaller, isolated sections (stages) of the well. Fracturing discrete intervals also allows operators to make adjustments for variations in the formation, such as shale thickness, the presence of natural fractures, and proximity to fractures from a nearby well.

Approximately 5 million gallons of water are required for each hydraulic fracturing job.[7] After a hydraulic fracturing job is completed, the formation pressure causes some of the water in the fracture fluid to come back out of the well (flowback water), initially at a high rate. Some of the sand grains remain in the rock fractures, propping them open and allowing the gas to move. Next, production tubing is lowered to the depth at which fluids have accumulated and the space between the tubing and the casing is sealed with an inflatable packer to ensure that fluids enter the tubing. With this step, the well has been completed and is ready for production.

In the production stage, natural gas, water, and any natural gas liquids flow from the formation into the well. The fluids are separated, and the gas is transported through pipelines to a gas-processing facility. The water, which is a blend of flowback water and produced water (water naturally occurring in the shale), is collected and then managed by underground injection, treatment and reuse, treatment and discharge, or evaporation.

[6] See http://lingo.cast.uark.edu/LINGOPUBLIC/natgas/index.htm, cited in John Veil's presentation.

[7] Presentation at the workshop by John Veil, Veil Environmental, LLC, on September 9, 2013.

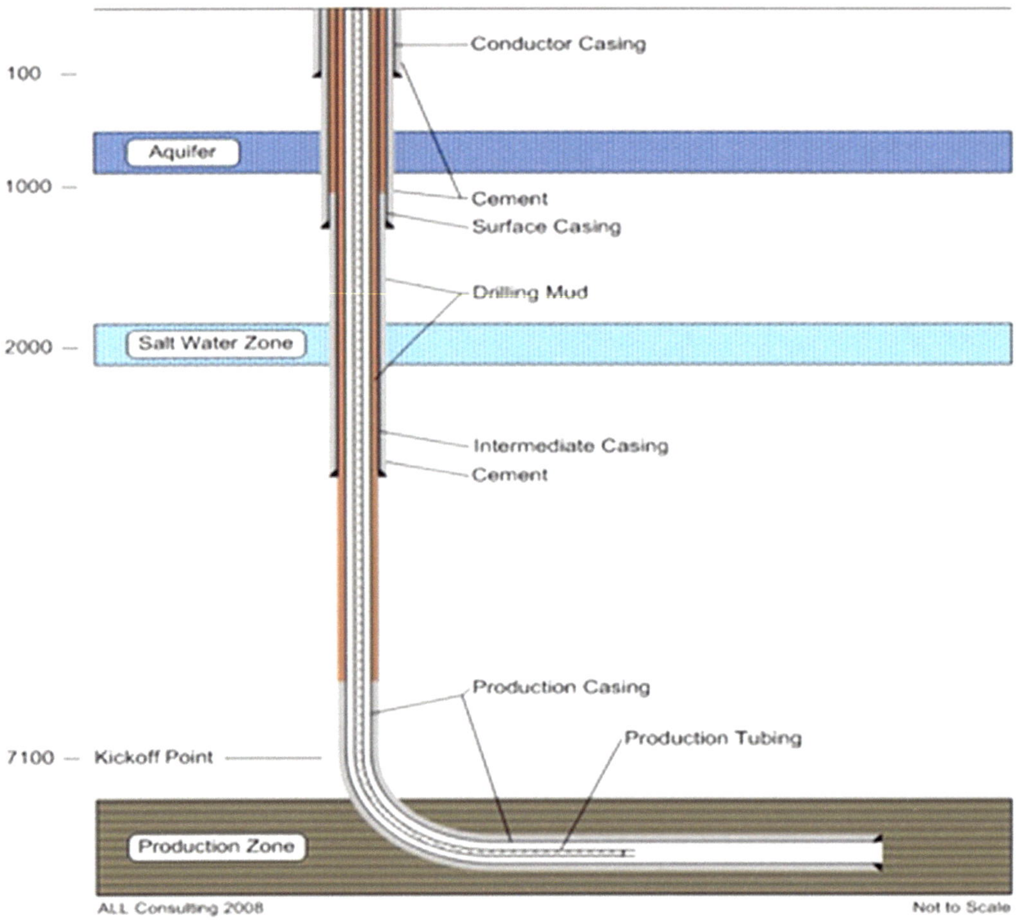

FIGURE 1.3 Schematic of the casing and cement installed in shale gas wells. This illustration shows four concentric strings of casing (conductor, surface, intermediate, and production) and a small-diameter production tubing string. The conductor casing prevents surface soil from caving into the well; the surface casing seals off freshwater aquifers; and intermediate casing seals off any saltwater zones. The depths (in feet) are illustrative. SOURCE: ALL Consulting, http://energyindepth.org/ohio/the-myths-and-realities-of-horizontal-drilling-and-fracing/.

ORGANIZATION OF THE REPORT

This report focuses on the geologic, environmental, and engineering aspects of unconventional hydrocarbon production in the Appalachian Basin. As such, it complements other recent NRC workshops that focus on health effects (IOM, 2013) and risk management and governance (NRC, in preparation). The organization of this report mirrors the structure of the workshop, which roughly followed the Statement of Task. Chapters 2, 3, and 4 summarize the plenary and working group presentations for the three sessions of the workshop. Chapter 2 covers the geology of the Appalachian Basin, unconventional resources and how they are produced, and the potential of these production activities to induce earthquakes (Tasks 1 and 4). Chapter 3 covers the impact of unconventional hydrocarbon production on water quality and water quantity (Tasks 2 and 4) and also summarizes federal and state regulations aimed at protecting water and other elements of the environment. Chapter 4 covers the potential impacts of production on ecosystems, air quality, and climate (Tasks 3 and 4). Finally, Chapter 5 summarizes closing remarks made by the planning committee and other workshop participants.

Supporting material for the report appears in the bibliography, which include papers cited in the presentations or discussed in the working groups, and the appendixes. A letter from John D. Rockefeller, U.S. Senator for West Virginia, is given in Appendix A. The workshop agenda and list of participants appear in Appendixes B and C, respectively. The reports made by the working groups are given in Appendix D. Biographical sketches of planning committee members appear in Appendix E. Finally, acronyms and abbreviations appear in Appendix F.

This report has been prepared by the workshop rapporteur as a factual summary of what occurred at the workshop. The planning committee's role was limited to planning and convening the workshop. The views contained in the report are those of individual workshop participants and do not necessarily represent the views of all workshop participants, the planning committee, or the National Research Council.

2

Geology, Resources, and Production

The first session of the workshop provided an overview of the geology and unconventional hydrocarbon resources in the Appalachian Basin, the technical and engineering processes used to produce unconventional resources, and the potential of these production activities to induce earthquakes. Following three plenary talks, workshop participants broke into working groups to discuss unconventional hydrocarbon resources, engineering practice, and induced seismicity, and to identify priorities for future research, as summarized below.

PLENARY PRESENTATIONS

Geology, Resources, and Potential Activity Levels

Ray Boswell, Department of Energy National Energy Technology Laboratory

Boswell noted that the origins of unconventional hydrocarbon resources in the Appalachian Basin date back to the early Paleozoic, hundreds of millions of years ago. At that time, the basin was covered by ocean and flanked by mountains to the east. Periodic mountain building cut off the basin from the global ocean, creating anoxic conditions that allowed organic matter to be buried in marine muds and eventually converted to hydrocarbons. Some of the hydrocarbons stayed in place as the mud was converted to shale through heat and pressure, and some migrated into overlying sand reservoirs.

These geological events created five main unconventional hydrocarbon plays in the north-central Appalachian Basin:

1. Utica–Point Pleasant shales, which produce dry gas in the east (western Pennsylvania and central New York), wet gas near the Pennsylvania–Ohio border, and oil in the west (eastern Ohio), reflecting an east-west temperature gradient reached during burial. The shales are amenable to fracture stimulation because they are low in clay and high in calcite or quartz. Production is in the early stages and is focused primarily in eastern Ohio.

2. Clinton–Medina basin-centered sands, which has produced naturally flowing or technologically stimulated gas for more than a century. More than 70,000 wells have been drilled, primarily in Ohio. Development of the play in New York, Pennsylvania, and West Virginia has been discouraged by the economics of production.

3. Marcellus Shale, which is a low-clay, high-quartz shale that produces dry gas in northeastern Pennsylvania and more liquid-prone gas in southwestern Pennsylvania and northern West Virginia (Figure 2.1). Approximately 8,000 wells have been drilled thus far, perhaps less than 10 percent of the total potential.

4. Ohio and other Upper Devonian shales, which have produced gas for more than a century in West Virginia and eastern Kentucky. These clay-rich, naturally fractured shales supplied one of the biggest gas fields in the United States in the 1930s.

5. Upper Devonian–Lower Mississippian basin-centered sands and silts, which produce gas from conventional reservoirs in favorable settings. Hundreds of thousands of wells have also been drilled in lower quality tight sands and silts. These formations began producing in 1859.

Boswell next summarized the hydrocarbon resources (see Box 2.1 for definitions) contained in these five plays and the methods used to assess them. He described two basic methods: (1) assessments of gas in place, which estimate how much gas exists in a play using geologic parameters (e.g., porosity, hydrocarbon saturation) and an engineering factor; and (2) assessments of techni-

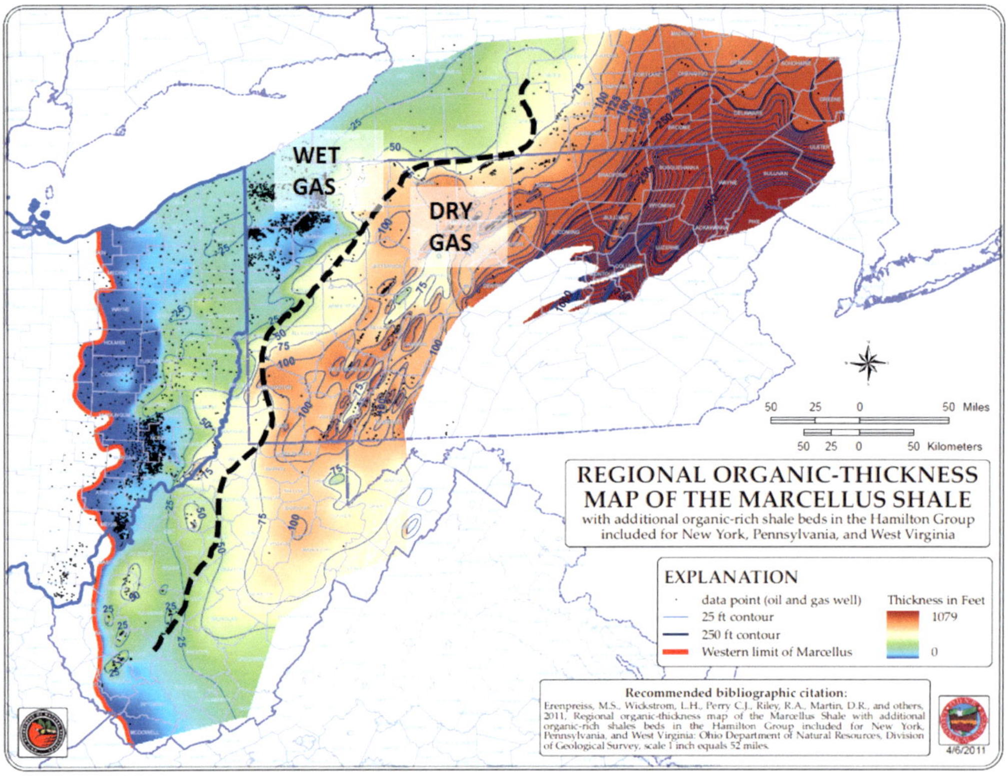

FIGURE 2.1 Thickness and extent of the Marcellus Shale. The wet gas–dry gas transition is marked by the dashed black line. SOURCE: Modified from Erenpreiss et al. (2011).

BOX 2.1
Definitions of Resources

In his presentation, Boswell distinguished four types of resources:

1. Gas in place—all gas that exists in a play, which is a function of the geology.
2. Technically recoverable resources—all gas that could be expected to be produced, which is a function of the geology as well as the technologies, policies, and regulations.
3. Economically recoverable resources—the subset of technically recoverable resources that can be produced at a profit, which is a function of market conditions.
4. Reserves—the economically recoverable resources that have been proven to exist by drilling and are available for economic production.

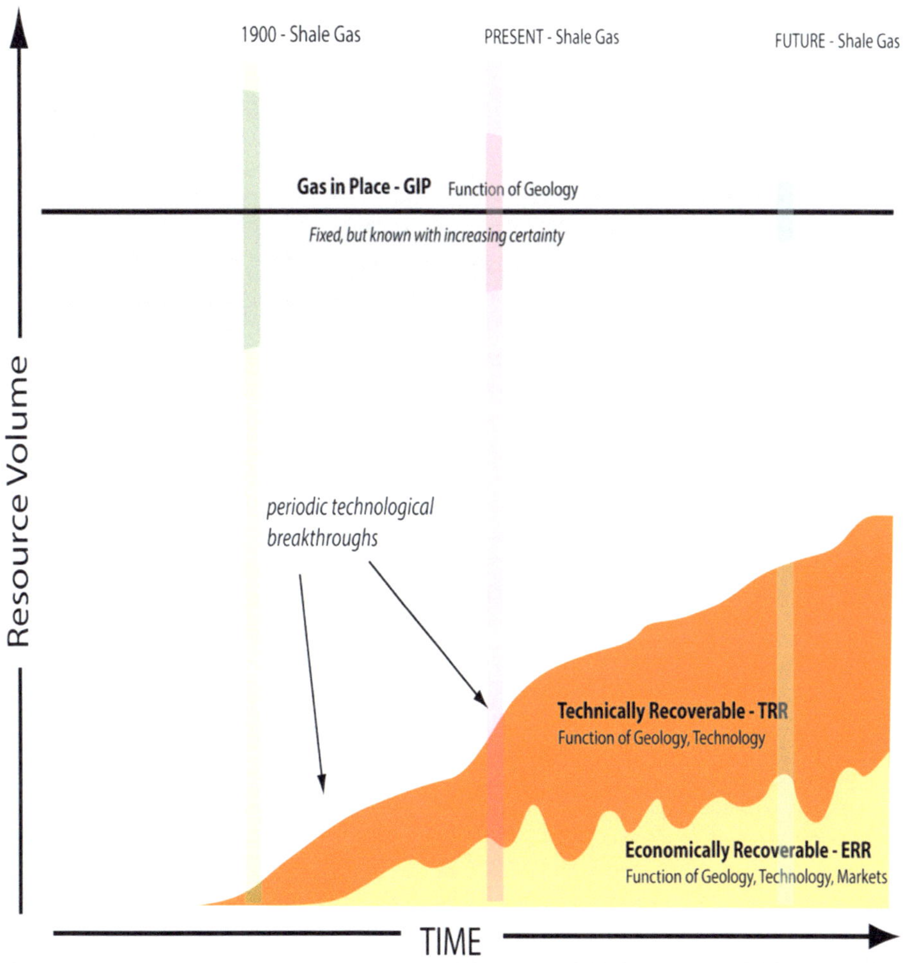

Volumes of different types of resources over time. Technological breakthroughs periodically increase the technically recoverable resources. Estimates of gas in place remain relatively constant, but uncertainties (gray shaded area) decrease with time. SOURCE: Ray Boswell, National Energy Technology Laboratory, from Boswell and Collett (2011). Reproduced by permission of The Royal Society of Chemistry.

cally recoverable resources, which estimate how much gas could be ultimately recovered from the play from historical production data for each well, the number of remaining well locations, and assumptions about the geology, market conditions, technology, and regulations. Assessments of technically recoverable resources are made periodically by federal government agencies such as the U.S. Geological Survey (USGS) and the Energy Information Administration (EIA). The assessments can change significantly over time because there is little production data in newly developing areas, and technological, economic, and regulatory factors evolve. For example, before the first Marcellus well was drilled, the USGS assessed a mean of approximately 2 trillion cubic feet (TCF) of gas for the Marcellus Shale (Milici et al., 2002). Less than a decade later, the Marcellus assessment had increased to a mean of 84 TCF of gas (Coleman et al., 2011).

Boswell concluded that, despite uncertainties, the Marcellus play has the potential to sustain production in the Appalachian Basin over several decades. In addition, a large number of wells could potentially be drilled, on the order of 2,000 per year for both the Marcellus and Utica shales. Detailed production and development data could be used to improve understanding of the resource and help yield more accurate assessments.

Questions. A workshop participant asked whether the U.S. government is working to fill gaps in knowledge about reserves in other countries, and Boswell answered that the USGS and the EIA have recently issued reports about global shale resources.[1] Another participant asked for the current average estimated ultimate recovery values for wells in the Marcellus and the Utica shales. Boswell deferred to Joseph Frantz, who guessed that they are 6–25 billion cubic feet (BCF) per well. Finally, a participant asked about the advantages and disadvantages to private companies for providing the detailed production and development data needed to improve estimates of reserves. Frantz said that each operator typically records a daily production number, which is used to forecast future well performance. Cumulative values are provided to Pennsylvania every 6 months and to many other states every month, as required by state law. These cumulative values mask what is happening on a daily basis, but he thought that operators would not object to providing the data more frequently if asked.

Engineering and Technology for Developing Unconventional Resources: Current and Prospective Methods for Exploration and Technology

Joseph Frantz, Jr., Range Resources

Frantz commented that early assessments indicate huge shale resources in China, Australia, and Argentina. Some of these countries are sending experts to the United States to learn how shale resources are being developed and to understand key field issues, such as the transport of water to the wells. With gas prices relatively low, U.S. operators are drilling where the gas contains natural gas liquids (e.g., propane, butane), which provide additional revenue.

In 2000, U.S. shales produced about 1 BCF of gas per day; today the Marcellus Shale is producing 11–12 BCF per day. Production increased dramatically in the late 2000s when the industry succeeded in isolating different intervals in a horizontal well and pumping multiple fracture treatments into that well. Other recent improvements include the following:

- Drilling multiple wells from one pad (pad drilling), which results in a smaller surface footprint of production than the cumulative footprint of the same number of vertical wells and associated roads.

[1] See http://pubs.usgs.gov/fs/2012/3042/ and http://www.eia.gov/analysis/studies/worldshalegas/.

- Using focused ion beam and scanning electron microscope techniques to characterize the pore structure of fine-grained shales, which enables areas with the highest porosity, permeability, and gas in place to be targeted.
- Combining three-dimensional seismic and microseismic techniques with treatment data and models to deduce what part of the fractured area is producing (effective fractured area). Increasing the effective fractured area is an ongoing research effort.
- Improving horizontal drilling techniques, such as using longer lateral wellbores and shorter stages, which increases production.
- Improving the efficiency of drilling and completion through technological advances or practice, which increases the number of stages that can be completed each day and the number of wells that can be drilled each year.

Frantz also touched on some safety and environmental measures undertaken by industry, including using multiple layers of casing, cemented back to the surface, through groundwater zones; placing rubber containment under every piece of equipment that could leak; using impoundments and temporary lines to store and transfer water to the wells, thereby reducing truck traffic; and improving facilities to capture fugitive emissions. Future improvements include further reducing the surface footprint of production and using less water.

These recent changes demonstrate the fast-paced nature of the industry, with ongoing development of new technologies and the constant introduction of new regulations that change industry practices every few years. Frantz concluded that natural gas production is a big opportunity for the United States and that industry must be diligent to maintain its social license to operate.

Questions. A workshop participant asked whether the microseismic data indicate which direction the fractures tend to propagate. Frantz said that hydraulic fractures have a propensity to propagate in the same direction. In the Marcellus, microseismic data and logs show conclusively that the fractures are going northeast-southwest. Another participant asked whether existing wells are being re-stimulated, and Frantz answered that the wells are still too productive to consider using that practice. When a well produces less than 500 million cubic feet of gas per day, the industry will likely look to stimulate the zones between the existing perforations.

A participant questioned whether pad drilling significantly reduces the number of vertical wells that would be needed, and Frantz answered that prior to the recent development of shale-gas reservoirs, most of Appalachia was drilled on about 40-acre spacing and less for economic plays. The participant asked if anything constrains where a well pad is placed or how many well pads can be placed on a landscape, and Frantz said that a small-footprint operation with a small rig requires a narrow access road and a small amount of land that is flat or can be leveled.

Finally, one participant wondered why a lining is used under the entire site. Frantz said that there are advantages and disadvantages (e.g., truck traffic) to lining the entire site, and thought it likely that a fair number of companies are lining only parts of the site.

Earthquakes Induced by Hydrocarbon Production: What Texas Can Tell Us About Appalachia

Cliff Frohlich, University of Texas, Austin

Frohlich noted that earthquakes associated with oil and gas production have been documented in several states, including Texas, Arkansas, Colorado, Ohio, and Oklahoma. His research has focused on earthquakes in Texas, but the lessons learned are applicable to the Appalachian Basin.

Texas has produced huge volumes of oil and gas for more than 80 years. Statistical analysis shows that earthquakes in Texas are sometimes caused by the disposal of waste fluids in injection wells and, in some areas, by the extraction of gas and fluids. (Earthquakes do not appear to be caused by drilling, and only a half-dozen cases of earthquakes induced by hydraulic fracturing have been documented.) A strong case can be made that earthquakes are induced by waste fluid injection when the earthquakes begin more than a year after injection commenced, the earthquakes are within 2–3 km of injection wells, the injection wells handle high volumes of water (>100,000 barrels per month), the earthquakes occur along previously unknown faults, and earthquake depths are at and below the depth of injection.

The vast majority of injection wells do not cause earthquakes, and the vast majority of injection-induced earthquakes are small (lower than magnitude 3 or 4). However, a few of these earthquakes are larger (magnitude 4 and 5), which can be a concern in populated areas. Why some injection wells cause earthquakes while others do not is poorly understood. It is possible that earthquakes occur when there are suitably oriented faults near an injection well. Perhaps fluid injection pushes the rock on each side of the fault apart, reducing friction and allowing slip. However, Frohlich added that more research is needed to understand why injection near a fault is sometimes a problem and sometimes is not.

Frohlich noted that both Texas and Appalachia are geologically diverse and have natural earthquakes. If unconventional hydrocarbon development proceeds in the Appalachian Basin over the next 50 years, then there will almost certainly be some induced earthquakes.

Questions. A workshop participant asked why some earthquakes occur below the injection point, and Frohlich said he was unsure; it is an empirical observation that faults tend to be at that depth. A participant asked if the fluids being injected are high density and whether that might account for the percolation of fluids to a greater depth. Frohlich deferred to Frantz, who said that most oilfield brines are fairly heavy.

Another participant asked whether it is possible to use reverse modeling to determine the capacity of the well and where it was stimulated by the injection. Frohlich said that careful modeling of the subsurface using detailed information about the injection, hydrology, and geology is only beginning to be carried out. However, he does not expect modeling to yield a definitive answer (e.g., earthquakes will not be induced if injection rates stay below 150,000 barrels of water per month) because the geology differs from place to place.

Finally, a participant asked whether regulators can do anything to prevent injection-induced earthquakes, given that the causal mechanism is unclear. Frohlich said that regulators should invest in some fairly stringent regulations for operations in urban areas, but regulations are not needed in areas where an earthquake would have little impact. Installing seismic networks around all injection wells is not affordable because there are too many wells and not enough earthquake seismologists to examine the data.

WORKING GROUP REPORTS

Each working group was asked to consider the following themes:

- Geology and hydrocarbon resources—Estimated resources of natural gas, oil, and/or natural gas liquids and current and projected production levels in the main hydrocarbon-bearing geologic formations in and around the Appalachian Basin (e.g., Marcellus, Utica, and Devonian shales);
- Technical and engineering processes—Current and prospective technical and engineering processes for exploration and production of hydrocarbons from unconventional resources; and

• Research priorities—Scientific and engineering research needed to narrow or characterize uncertainties.

Issues that were directly related to these themes or that were raised by more than one working group are summarized below. The complete working group reports appear in Appendix D.

Research and Development Related to Geology and Hydrocarbon Resources

Research and development needs identified in working group discussions included the following:

• Characterizing the shale formations, including porosity, permeability, and spatial variations;
• Improving understanding of the controls (e.g., geologic factors, reservoir pressure) on induced seismicity;
• Improving understanding of the geologic controls (depositional environment, structure, thermal history) on the productive, operational, and economic lifetimes of wells in different areas of the shale plays;
• Developing ways to calculate stimulated rock volume and to relate it to gas in place, estimated ultimate recovery, and resources and reserves; and
• Determining the relationship between well completion strategies and estimated ultimate recovery.

Monthly and daily production data would facilitate this research and improve estimates of resources and reserves for a region, field, or individual well. In addition, seismic measurements and cores, especially around high-rate injection wells, are needed to improve understanding of induced seismicity. Some of these data exist, but are proprietary (e.g., monthly industry data) or are not readily accessible (e.g., some data collected by states). Some workshop participants thought that the shortage of public data adversely affects further development of shale resources and erodes the public trust.

Collaboration

Many of the issues associated with the geology and production of unconventional hydrocarbon resources—such as assessing the regional and cumulative economic, environmental, and social impacts—would benefit from a multidisciplinary, multisector approach. However, it is a challenge to balance industry's immediate needs with the comparatively slow pace of research and the time it takes to establish relationships with individual researchers, universities, or research consortia. Some workshop participants suggested creating a common vision for developing shale gas (e.g., number of wells, expected revenue) or using common terminology to facilitate collaboration across academic, government, and industry sectors. Others suggested that industry and state regulators work together to determine what data should remain proprietary and what data could be made public.

Communication and Education

Better communication with the public was a common theme of the working groups, although it was not always clear what message to send or how to send it. Some participants suggested that citizens, university scientists, and government scientists, managers, and regulators need to understand hydraulic fracturing as well as industry does. Some thought it would be useful to involve public policy makers and educators in the discussion. Others saw a role for universities to play an honest broker by providing credible information to the public on the costs and risks of shale gas development.

3

Water and Regulations

The second session of the workshop focused on methods for managing and disposing of wastewater from the production of unconventional hydrocarbons, the potential impacts of hydraulic fracturing and other production processes on water quality, and regulations aimed at minimizing the risk of adverse effects from unconventional hydrocarbon production on water and the environment. Following three plenary talks, workshop participants broke into working groups to discuss these topics and to identify priorities for future research, as summarized below.

PLENARY PRESENTATIONS

Water Issues Relating to Unconventional Oil and Gas Production

John Veil, Veil Environmental, LLC

Veil gave a brief review of the shale gas production process (see Chapter 1), noting that water plays a role in several steps, including drilling the well (e.g., lubricating the drill bit with drilling fluid or mud), preparing a well for production (e.g., hydraulic fracturing, collecting flowback water), and producing gas (e.g., collecting, treating, or disposing of fluids). Each hydraulic fracturing job requires about 5 million gallons of water, which comes from local ponds or streams, constructed reservoirs, the public water supply (e.g., used municipal water, groundwater), or wastewater from gas wells.

The amount of water used in shale gas production receives significant public attention. Veil calculated that about 6 percent of the wastewater generated by oil and gas production in the United States (1 million active wells) comes from shale gas production (20,000 wells). Veil also compared the amount of water required for hydraulic fracturing of the Marcellus Shale with the amount of water used for other purposes. He calculated that a high production year for the Marcellus may require 80 million gallons of water per day in New York, Pennsylvania, and West Virginia,[1] compared with 17,120 million gallons per day withdrawn by the thermoelectric industry and 24,577 million gallons per day for all purposes in the three states (Kenny et al., 2009).

The management of wastewater is complicated by rapid changes in water quality and water quantity over time. The volume of flowback water declines sharply after the first few hours to days while concentrations of total dissolved solids and other constituents in flowback and produced water increase rapidly. The most common way to manage wastewater is to inject it into a disposal well. Other management methods include removing metals and other contaminants to create clean brine; desalinizing clean brine to create clean freshwater; evaporating the water to dryness or crystalline form; and filtering flowback water to remove suspended solids and blending it with freshwater for use in a subsequent fracturing job. Because only 10–20 percent of the volume of fracture fluid returns to the surface, 80–90 percent freshwater must be added to the flowback water for recycling.

In Pennsylvania, a substantial amount of the flowback and produced water is being reused, in part because options for nearby injection wells are limited. At some point in the future, the number of new wells being drilled will decline while all the old wells continue to produce water. When more water is produced than can be recycled, additional water management facilities will be needed in the region.

Veil closed by identifying some water-related practices that are relatively environmentally friendly, including siting wells away from streams and water wells, collecting water samples before drilling to have a baseline for assessing contamination, constructing wells using appropriate materials, testing to ensure well integrity, air drilling the top section of wells to reduce water needs, using lined reserve pits or tanks for drilling wastes, reusing wastewater, and emphasizing piping over trucking of water.

Questions. A workshop participant asked why so much fluid stays underground, and Veil answered that the fracture process creates significant surface area, which attracts moisture. The rock actually imbibes some of the water. Shales like the Barnett return more water than is put in the well, whereas shales like the Marcellus return only a small fraction of water that is put in.

Identifying and Assessing Potential Impacts of Unconventional Hydrocarbon Production on Surface and Groundwater Quality

Rosemary Capo, University of Pittsburgh

Capo noted that the risks to groundwater quality vary with the production activity. For example, a loss of well integrity when surface casing is being set can allow fluids to migrate from the open hole, and chemicals stored at the surface for well completion and hydraulic fracturing can spill or leak. Many of these risks can be managed through due diligence in the shale gas development process.

The primary water quality issues related to unconventional gas production in the Appalachian Basin include high concentrations of total dissolved solids in the produced water, the migration of stray gas into the water supply, and the potential migration of water from deep formations into shallow aquifers. Produced water from the Marcellus Shale is high in total dissolved solids, typically 180,000 milligrams per liter after a few weeks of production, posing challenges for recycling and

[1] This hypothetical maximum was estimated by doubling the highest number of wells drilled in Pennsylvania, West Virginia, and New York and multiplying by 5 million gallons of water per well. The number of wells in Pennsylvania and West Virginia were based on state records. For this calculation, the number of wells in New York (currently zero) was assumed to be the same as the number of wells in West Virginia (922).

for wastewater treatment. Before the produced water can be used in subsequent fracture jobs, the fracture fluid mix must be adjusted substantially to avoid degrading well performance.

In addition, high concentrations of particular chemical constituents can be problematic. For example, barium can combine with sulfate in the water used for recycling and cause sediment to build up in the wells. Potentially hazardous species such as uranium and radium can mobilize from drill cuttings under acidic conditions. Although uranium is low in water produced from the Marcellus Shale, the concentrations of radium and strontium tend to be high and to increase over time.

Marcellus-produced water is also high in bromine, which can react with organic compounds in surface water to produce trihalomethanes, a possible health risk. Elevated levels of bromine have been detected in surface water near Pittsburgh, and studies are under way to determine whether the source is industrial wastewater treatment plants, which treat fluids from oil and natural gas production, or coal-fired power plants.

Combustible gas such as methane is common in water supply wells in the Appalachian Basin. Potential sources include naturally occurring gas seeps; abandoned or operating gas wells, coal mines, or landfills; coalbed methane wells; natural gas storage fields and pipelines; shallow formations and aquifers; buried organic matter; drift gas deposits; or other sources. Some sources of methane (natural gas or coal beds, landfills, and drift gas deposits) can be distinguished using stable isotope measurements, although the data can be difficult to interpret if the gas is a mixture from multiple sources. Better techniques for analyzing methane isotopes and more baseline measurements of water quality before and after drilling would improve identification of the source(s) of stray combustible gas.

The extent to which water from deep formations can migrate into shallow aquifers is unclear. The high total dissolved solid content of some artesian water suggests that water from deep formations can migrate along natural fractures into shallow aquifers, although this interpretation has been questioned. Some model simulations have suggested that hydraulic fracturing can create conduits for brine migration above the target formation, although the modeling methodology has been challenged. Strontium isotopes may prove useful for tracing the movement of fluids in the environment because Marcellus-produced waters have a different isotopic composition than other sources, including conventional Upper Devonian gas reservoirs, abandoned coal mine draining, fly ash impoundments, and coal-fired power plant discharges (Figure 3.1). Early results (Capo et al., in press) show that even small amounts (<0.1 percent) of Marcellus-produced water in groundwater or surface water would significantly shift strontium isotope ratios.

Capo concluded that water quality studies in the Appalachian Basin are complicated by the geologic variability and the legacy of coal and mineral mining, oil and gas production, and industrial activities, all of which can contribute contaminants. Developing the baselines needed to track these sources will take significant time and effort.

Questions. A workshop participant said that the Pennsylvania Department of Environmental Protection seems unwilling to disclose whether bromide levels in the Allegheny River have changed. If they have not changed, then perhaps shale gas production is not the source. Capo said that daily sampling results published recently show spikes in bromide levels, some of which exceed levels recorded over the past 5 years. It appears that only one nearby centralized wastewater treatment facility has treated Marcellus-type fluids, but coal-fired power plants in the area use brominated solvents. Her group is sampling individual creeks and streams near these treatment centers and power plants to try to resolve the source(s).

Another participant asked how uranium and radium are removed from water and what is done with the residual material. Capo said that uranium tends to stick to particles, so the uranium content of produced water is low even if the uranium content of organic-rich black shale is high. Some people have proposed trying to recover uranium from drill cuttings before the cuttings are disposed

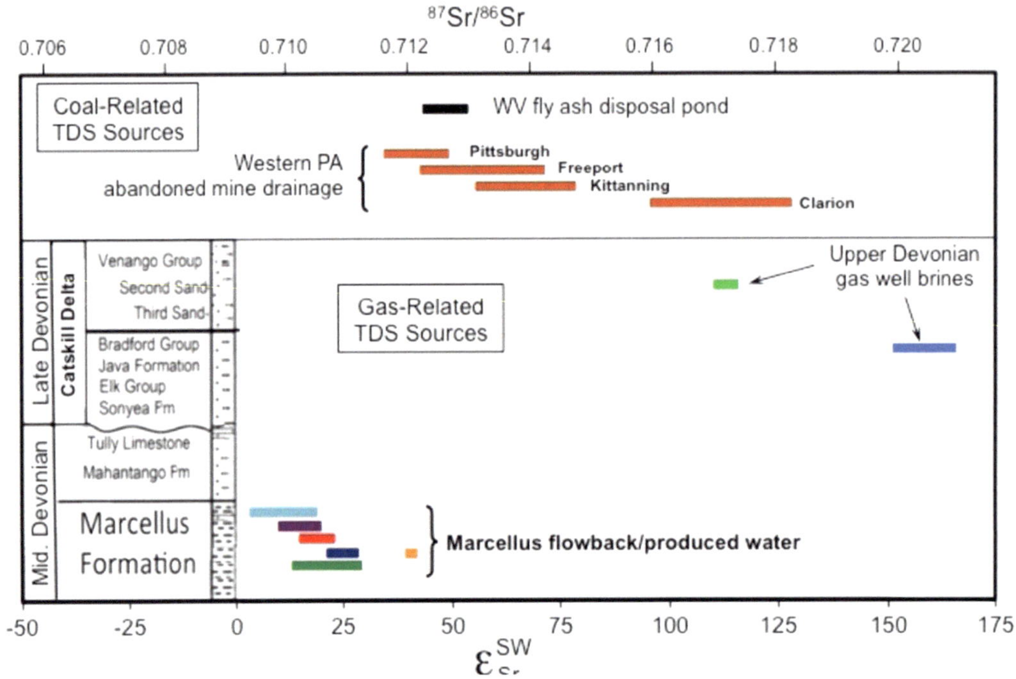

FIGURE 3.1 Strontium isotopic signature of water from shale gas production (*bottom*) compared with the isotopic signatures of coal-related sources (*top*). SOURCE: Reprinted from Capo et al. (in press), with permission from Elsevier.

in landfills. Wiseman said that regulatory changes would be required to make that happen. If a regulation specifies disposal of waste in a particular manner, then it generally has to be modified to allow for reuse or recycling.

State Shale Gas Regulation in the Appalachian Basin: Recent Enhancements, Remaining Gaps, and Opportunities for Change

Hannah Wiseman, Florida State University

Wiseman noted that a large number of government regulations and enforcement strategies are aimed at protecting soil, surface water, and groundwater quality during well site activities. Relevant federal laws include the Clean Water Act, which prohibits the discharge of wastewater into streams and requires erosion control measures to prevent contaminated runoff from leaving the site during storms. In addition, most states have regulations governing the storage of flowback and produced water (e.g., requirements for pit lining and freeboard, well and site setbacks), well construction (e.g., well casing standards, blowout prevention equipment), and spills of onsite material (e.g., prevention and control plans, secondary containment requirements, erosion and sediment control plans). More states are beginning to require baseline testing, and West Virginia and Pennsylvania provide additional incentives for testing by presuming that contamination of water wells within a certain time and distance from hydrocarbon production is caused by the oil and gas operator.

The impact of wastewater disposal on soil, surface water, and groundwater quality is regulated by the federal Safe Drinking Water Act, which requires that disposal wells not endanger underground sources of drinking water. States implement this act for underground injection control wells, and Ohio also requires seismic testing and monitoring of the wells. Pennsylvania has few underground injection control wells and requires a waste management plan and reuse strategies. The Environmental Protection Agency is drafting standards for treating wastewater.

Some states regulate the quantity of surface water withdrawn for fracturing by requiring permits or water management plans showing that aquatic resources will not be adversely affected. For example, West Virginia requires operators to demonstrate that streamflow will be sufficient downstream of the withdrawal and that adverse impacts on aquatic life will be minimized. Some multistate compacts require water withdrawal permits or set minimum flow requirements across watersheds.

Compared with the number of regulations focused on water, relatively few regulations focus on air quality or ecosystems. Air quality regulation is dominantly at the federal level. The Clean Air Act sets emission standards and covers some equipment for controlling conventional and hazardous air pollutants and for capturing volatile organic compounds and methane. Some states also require minor source permitting for conventional pollutants from wells, storage tanks, or equipment at sites, and some states have rules limiting venting, flaring, and gas leaks.

Only a few regulations cover habitat fragmentation or impacts to wildlife and landscapes. For example, the U.S. Fish and Wildlife Service has listed the diamond darter, a freshwater fish, as an endangered species and has designated West Virginia's Elk River as critical habitat. Maryland is proposing comprehensive drilling plans to minimize the area of surface disturbance and to prohibit well pads in certain sensitive areas.

In recent years, environmental protections have grown stronger, with states such as West Virginia, Pennsylvania, and Ohio expanding setbacks. Ohio has also added a baseline water testing requirement. Casing requirements are also being updated or expanded. In Ohio, for example, casing material must now comply with industry standards, and the casing must be installed in a manner that isolates underground sources of drinking water and hydrogen sulfide gas. Air quality protections have been strengthened in West Virginia, which is encouraging green completion technologies,[2] and in Pennsylvania, which is trying to reduce air emissions through leak detection, equipment repair, and other measures.

Of course, regulations are useful only if they are enforced. Enforcement requires qualified inspectors who know how drilling and fracturing work and what to look for at a site, and who are capable of identifying problems and violations in the many regulations governing unconventional hydrocarbon production. A number of states have established minimum qualifications for inspectors to encourage good enforcement and minimum salaries to encourage retention. For example, West Virginia recently set minimum salaries for inspectors ($35,000) and supervising inspectors ($40,000). Other approaches for improving inspection and enforcement include focusing on the highest-priority regulatory problems, establishing uniform inspection reporting to produce consistent information about problems (Pennsylvania), and disclosing violations and enforcement actions to the public (Pennsylvania and Ohio).

To fund enforcement, some states are increasing fees and penalties. For example, both West Virginia and Ohio have increased well permit fees and Ohio has increased waste injection well fees, particularly for waste produced in other states. West Virginia and Pennsylvania have increased penalties for violating oil and gas rules.

Wiseman concluded by identifying gaps in regulation and enforcement, including uniform collection and reporting of baseline testing data, bonds set high enough to cover site restoration,

[2] See Department of Environmental Protection New Source Performance Standard 0000.

environmental liability insurance, training of inspectors, and consideration of how biological and other processes may change contaminants in flowback water being stored for recycling.

Questions. One workshop participant asked Wiseman to comment on the validity of using the State Review of Oil & Natural Gas Environmental Regulations (STRONGER)[3] process to review state regulations as well as the direction the program is taking. Wiseman noted that STRONGER is a collaboration of industry, environmental groups, and state regulators. It arose in part because the Environmental Protection Agency decided against regulating most oil and gas exploration and production waste under the Resource Conservation and Recovery Act, and left it to the states to address any potential gaps in the management, handling, and disposal of waste. STRONGER has developed guidelines for good regulation, not only for handling waste, but also for other aspects of oil and gas development, including hydraulic fracturing. It reviews state regulatory programs for compliance with these guidelines, looking at both the substance of state regulations and enforcement (e.g., number of staff members). Wiseman thought that this process is crucial and that efforts to compare state programs and find gaps should be expanded. The primary weaknesses are that state participation is voluntary and that the extent to which states have implemented the STRONGER recommendations varies.

WORKING GROUP REPORTS

Each working group was asked to consider the following themes:

- Potential effects on surface water and groundwater quality and quantity—Connections between hydraulic fracturing and other production technologies and processes, and water systems, including scientific data and methods in assessing impacts;
- Technical and engineering processes—Use of water in producing hydrocarbons from unconventional resources and methods for managing and disposing wastewater; and
- Research priorities—Scientific and engineering research needed to narrow or characterize uncertainties.

Issues that were directly related to these themes or that were raised by more than one working group, including working groups in other sessions, are summarized below. The complete working group reports appear in Appendix D.

Research and Data on Surface Water and Groundwater Quality

The working group discussions covered water use, reuse, treatment, disposal, geochemistry, and their linkages. Research priorities identified by the working groups included assessments of the impact of chemicals or materials used in the preproduction or production stages on water quality, the variability of stray methane in well water, the fate of fluids that remain in the subsurface, and potential problems with naturally occurring radioactive material in wastewater and solid waste. Longitudinal studies on long-term processes and the impacts of shale gas production on water quality would yield information that could help the public differentiate between perceived and actual risk.

Every working group identified the importance of baseline data and ongoing measurements to assess and monitor water quality and fluid migration. Establishing a baseline would require decisions on what constitutes the baseline, which parameters to measure, and protocols for measuring

[3] See http://www.strongerinc.org/.

these parameters in consistent ways. Data collected for monitoring could be gathered into a public database or library to support research and communication with the public. Citizens could also be enlisted in collecting data for monitoring.

Engineering Practices to Reduce Water Volume and Impacts to Water Quality

Several working groups pointed out that the shale gas industry has developed best practices (related to casings, cement, and well completion) to minimize surface water and shallow groundwater contamination. In addition, the working groups had several suggestions for improving wellbore integrity, such as developing better cements, identifying methods for detecting leaks, and agreeing on standards, tools, and protocols for testing wellbore integrity. Maintaining the integrity of old and abandoned wells is a challenge, and one working group suggested developing sensors to monitor the infrastructure and, if possible, fluid composition at different depths.

Other technical and engineering issues raised in the working groups included reducing the volume of water for hydraulic fracturing and developing advanced water treatments, such as high-rate, in situ treatment techniques for flowback and produced water, and desalination procedures to improve water quality and to recover valuable compounds. Disposal of produced water and cuttings, some of which contain radioactive materials, remains an issue. Simply keeping up with numerous and changing practices in wastewater treatment, disposal, transport, reuse, injection, and subsurface transport is a challenge.

Regulations

The working groups observed that regulations and standards as well as their application and enforcement vary across jurisdictional boundaries. As a result, it can be a challenge for companies to operate and for regulators to share data across jurisdictions. Another issue is that resources and inspectors are insufficient to guarantee regulatory compliance. Some suggested developing a process for updating state and multistate regulations and standards, and some suggested developing standards, criteria, and metrics for inspectors.

Working group participants disagreed on whether there are too many or too few regulations. Regulations generally cover specific practices, and rapid changes in water management practices mean that new regulations are continually being written. Some suggested targeting any new regulations to address vulnerable points and gaps in water management. Others suggested finding ways to develop a few good regulations that address the major environmental concerns.

4

Ecosystems, Air, and Climate

The third session of the workshop examined the potential impacts of unconventional hydrocarbon production on aquatic and terrestrial ecosystems, air quality, and climate. Four plenary talks were followed by working group discussions, which focused on potential impacts, methods for reducing these impacts, and priorities for future research, as summarized below.

PLENARY PRESENTATIONS

Potential Impacts of Unconventional Hydrocarbon Production on Stream Biota: Current and Needed Research

Kelly Maloney, U.S. Geological Survey

Maloney noted that little data have been collected on the direct effects of hydrocarbon production on stream biota, and so inferences have to be drawn from research on other types of disturbances to landscape conditions, such as agriculture, urbanization, and road construction. Stream communities—including primary producers (e.g., algae, plants), benthic microinvertebrates (e.g., insects, crayfish), fish, amphibians and reptiles, and birds and mammals—are sensitive to these disturbances and have been studied for decades to assess stream health.

Stream ecosystems may be affected by unconventional hydrocarbon production, particularly site preparation, which disturbs the land surface, and the use, treatment, and disposal of water. Recent research has highlighted three key potential issues for stream biota: habitat fragmentation, stream flow alteration, and degradation of water quality. Habitat fragmentation arises from the installation of roads and pipelines, which bisect landscapes, and from the loss of land through development of well sites. The Nature Conservancy (Johnson, 2010) estimated that the production infrastructure (roads, pipelines, well pads, impoundments) disturbs about 9 acres per well pad. Habitat fragmentation affects species in several ways. Smaller patches of habitat have lower diversity, and when patches of habitat become isolated from one another, it becomes more difficult for species to interact. The net effect is lower recolonization, reduced population sizes, and genetic bottlenecks. A fragmented habitat also has more edges, which benefits some species and harms others.

Water used in the hydrocarbon production process is often taken from streams and rivers, reducing or altering stream flows in ways that may adversely affect stream biota. For example, withdrawals that reduce pools below a threshold size at key times (e.g., spawning, incubation) can result in habitat loss. Temporary dams constructed to impound water (cofferdams) can change the stream conditions from lotic (flowing water) to lentic (still water), altering the habitat. Withdrawals of water that expose substrate can desiccate taxa that cannot move and also isolate pools, potentially stranding species and raising water temperatures above levels tolerated by some species. In a worst-case scenario, water withdrawals could alter the natural flow regime (quantity, timing, and variability of stream flow) on which stream organisms depend, negatively affecting instream biota.

Stream water quality may be degraded by contamination from spills and by sedimentation. Of particular concern are contaminants that increase stream salinity or decrease pH, which control the distribution of species in aquatic habitats, and heavy metals, which may accumulate in tissues, affecting physiology, growth, behavior, and reproduction. Research documenting sediment runoff from well pads shows a positive correlation between well density and stream turbidity. Sedimentation in streams can result in a loss of habitat or a loss of sensitive species. The extra sediment carried by streams may bury primary producers such as algae or scrape them off rocks, or it may fill interstitial habitats favored by benthic microinvertebrates or overwhelm the ability of these organisms to filter feed or breathe. Impacts to fish include reduced foraging efficiency, loss of pool and spawning habitat, and increased mortality of eggs laid in interstitial areas as a result of oxygen deprivation or sediment burial. Sediment also coats egg masses of amphibians and reptiles and may lead to a loss of sensitive species.

A few programs are beginning to be developed to restore habitats affected by the energy industry. For example, Wildlife Incentives for Nongame and Game Species (WINGS) is an effort by local government agencies, land trusts, and energy companies to create or enhance wildlife habitats along natural gas pipeline and electricity corridors.[1]

Maloney concluded with some topics for discussion, including the extent to which the wealth of data from activities such as agriculture and construction can be used to guide management and research on unconventional hydrocarbon production. Other issues include the availability of indicator species or biomarkers to detect habitat disturbance, the effects of habitat fragmentation and invasive species on stream ecosystems, and the efficacy of industry best management practices.

Questions. A workshop participant noted that short-term impacts can be measured (e.g., number of fish killed from a chemical spill). How can long-term impacts be measured? Maloney said that events such as a spill can be hard to detect because they move through the system quickly. Real-time monitors that collect data routinely are needed to detect a stress or mortality event. Monitoring and research are also needed to assess longer-term impacts, such as those caused by sedimentation. Sediment can be held in reservoirs for decades, and so the impacts may lag the activity that produced erosion and runoff.

The same participant observed that research, monitoring, and impact analysis cost money. Work on acute events is difficult to do without timely access to funds, and funding for water monitoring is declining. Some participants added that funding for stream flow measurements has been declining since about 1980. Moreover, proposals that include monitoring are not well reviewed by peers or by National Science Foundation program managers. Other participants thought that government and academic researchers are always going to be behind the curve because of the difficulty of raising funding, the time it takes to analyze results, and the dynamic nature of the industry.

[1] See http://www.tworiversrcd.org/index.php/redshop/wings.

Several participants suggested ways to stretch funding or partner with industry. Maloney thought that more use could be made of existing data sets, such as those collected by state agencies. Other participants suggested university or government collaborations with industry to obtain industry funding, data, or insights. Sustained collaboration could generate the knowledge and data sets needed to answer key questions and monitor environmental impacts.

Assessing and Minimizing Ecological Impacts of Shale Development

Michael Powelson, The Nature Conservancy

Powelson discussed the potential environmental footprint of unconventional hydrocarbon development. The Nature Conservancy develops ecological scenarios using information on reserves and current trends in energy development to create models of future development patterns. Ecological data on intact habitats, species distribution, migration patterns, and climate resilience are then integrated into the models to project potential long-term ecological impacts. Powelson showed several example projections. Under a medium-development scenario (10,000 new well pads by 2030, about 60 percent in forested areas), most of the intact forest blocks in central Pennsylvania would be eliminated or fragmented by 2030 (Figure 4.1). This is important because intact forest blocks are the critical resource that maintains biodiversity. A high level of development (15,000 new well pads by 2030) would affect more than half of the habitat occupied by brook trout and by black-throated blue warblers, two potentially endangered species.

Powelson emphasized that certain habitats, species, and ecosystems provide crucial ecological value to larger systems and landscapes. Development in these areas could reduce the viability and resilience of the entire ecosystem or landscape. He thought that avoiding development may be the

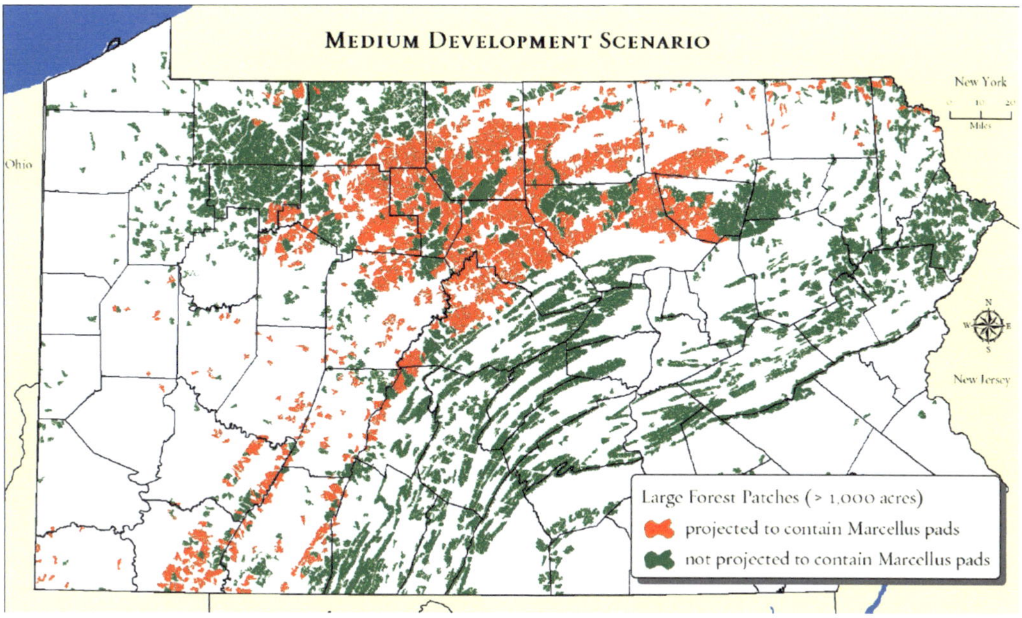

FIGURE 4.1 Projected forest fragmentation in 2030, assuming 10,000 new well pads with an average of 6 wells per pad. SOURCE: The Nature Conservancy.

most effective way to minimize ecological impacts. When avoidance is not feasible, steps can be taken to minimize impacts (e.g., by collocating roads and pipelines) or to offset impacts (e.g., by restoring affected areas). Both strategies would be facilitated by landscape-level planning, which examines cumulative effects across a broad geographical region and conservation priorities, as well as investigation and monitoring of ecological impacts.

Questions. A workshop participant said that many environmental advocacy groups are trying to articulate best management practices for industry, but ongoing technological developments cause these best practices to constantly evolve. What is the current landscape of best management practices, and are industry players sharing their most recent developments? Powelson agreed that there are many sets of best management practices, including some developed in partnership with industry (e.g., practices for minimizing methane emissions and impacts on water quality developed for the Marcellus Shale by the Environmental Defense Fund, Shell, Chevron, and QET). Most of the industry is trying hard to minimize the environmental footprint; the problem is that there are some bad actors.

Assessing Emissions of Hydrocarbons from Rural and Natural Gas Drilling Impacted Areas in Pennsylvania

Jose Fuentes, Pennsylvania State University

Fuentes noted that hydrocarbons enter the atmosphere from several sources, including active and abandoned wells, storage tanks, and pipelines. The heavy machinery and vehicles used in the gas production process also release compounds into the air. These hydrocarbons remain in the atmosphere only in trace amounts, but they react with constituents such as hydroxyl radicals and thus can influence the oxidation capacity of the atmosphere. The reaction products of these gases are precursors to pollutants such as carbon monoxide and ozone. They also condense readily in the atmosphere and form aerosols, which affect cloud formation and regional climate. Finally, elevated concentrations of hydrocarbon species such as benzene, toluene, ethylbenzene, and xylene (BTEX) can adversely affect human health.

Fuentes's research team has been taking air samples across Pennsylvania to determine whether activities associated with natural gas production have a measurable impact on regional levels of hydrocarbons in the atmosphere. Samples were taken from rural areas to establish a baseline, from state parks to examine diurnal variability in biogenic hydrocarbons (i.e., those produced by trees in the daytime), and from areas near different types of gas production activities (drilling, flaring, operating for different periods). The results show that BTEX levels are very low in mostly forested areas. In areas with a high density of wells, the atmosphere contains more anthropogenically emitted hydrocarbons and fewer biogenic hydrocarbons.

Approximately 120 chemical species of hydrocarbons were found in the Pennsylvania air samples. For the most common of these species, the alkenes, concentrations were similar in all settings: native, forested, agricultural, gas well-impacted, and urban. BTEX was higher in urban settings than other settings, likely because cities have many sources of toluene, benzene, and xylene. Finally, isoprene and other gases that come from forests are among the hydrocarbon species with the greatest potential to affect the chemistry of the atmosphere through reactions with hydroxyl radicals.

Fuentes also showed some preliminary results from air samples taken at a farm. Passing vehicles create spikes in nitrogen oxides, which affect ozone formation and air quality. Steps that could be taken to understand the interaction between nitrogen oxides and hydrocarbons in the Appalachian region include establishing an air sampling network to identify species and sources of hydrocarbons and developing numerical models to calculate emissions. A baseline of observations would also help industry determine when and where leaks of methane occur.

Questions. A participant asked how to reconcile top-down and bottom-up measurements when designing a monitoring network. Fuentes said that both approaches are necessary. For reactive gases, monitoring stations must be close to the wells because some of those gases are short-lived. For long-lived gases such as methane, inferential (top-down) methods can be used to calculate the source or the strength of the leaks.

Climate Impacts of Shale Gas

Paulina Jaramillo, Carnegie Mellon University

Jaramillo discussed the climate impacts of unconventional shale gas production from a life-cycle perspective, which tallies impacts from preproduction (well-pad preparation, well drilling, hydraulic fracturing, well completion), production, processing, transmission and distribution, and combustion of the shale gas. Sources of emissions include fuel use, flaring and venting, methane leaks throughout the system, water consumption, pad construction, vegetation clearing, and the production of drilling mud and additives. Studies (e.g., Jiang et al., 2001) show that combustion is the dominant source of emissions over the life cycle of shale gas. The main sources of emissions from preproduction through distribution depend on the production rate and lifetime of the well. For a well that produces 3 million cubic feet of gas per day for 25 years, emissions are dominated by the production stage. For a well that produces only 0.3 million cubic feet of gas per day and lasts only 5 years, a significant fraction of emissions comes from the preproduction stage.

The data used to estimate emissions have a variety of uncertainties. The results of an uncertainty and variability analysis of life-cycle greenhouse gas emissions from natural gas are shown in the probability distribution in Figure 4.2. Superimposed on this distribution are life-cycle emissions estimates from other published studies (colored dots). Most of these estimates fall within the 90 percent confidence interval (62–72 g CO_2e/MJ) of the probability distribution, although one (Howarth et al., 2011) is significantly higher, in part because the analysis assumed a higher loss of natural gas throughout the life cycle of the system.

The effect of natural gas on climate depends not only on how much methane and carbon dioxide is going into the atmosphere (discussed above), but also on how the gas is used (e.g., electricity generation, industrial and home heating, transportation). Gas use has many elements. Emissions associated with electricity generation, for example, depend on the relative efficiencies of natural gas and coal power plants, how power plants are scheduled to produce energy (which depends on the marginal cost of production, technological constraints, and the need to instantly match supply and demand), plans to retire coal plants, and how renewable energy sources, which produce variable and intermittent energy, are integrated into the power system. Emissions from electricity generation can be reduced by replacing coal with natural gas because emissions from natural gas are lower than those for coal, and new natural gas power plants are significantly more efficient than coal plants. However, the reductions depend partly on the price of natural gas. A recent study (Venkatesh et al., 2012) found that the maximum reduction in emissions from the power system is about 15 percent when the life cycle of the fuels is considered, and much less if the price of natural gas rises above $3.5 per million BTUs (British thermal units).

These factors add considerable complexity to the analysis of climate impacts from natural gas. Climate impacts are generally assessed using climate models, but complexities in emission sources and energy uses are only beginning to be incorporated in models. For example, a recent study (Wigley, 2011) used a simplified climate model to examine how replacing coal with natural gas at

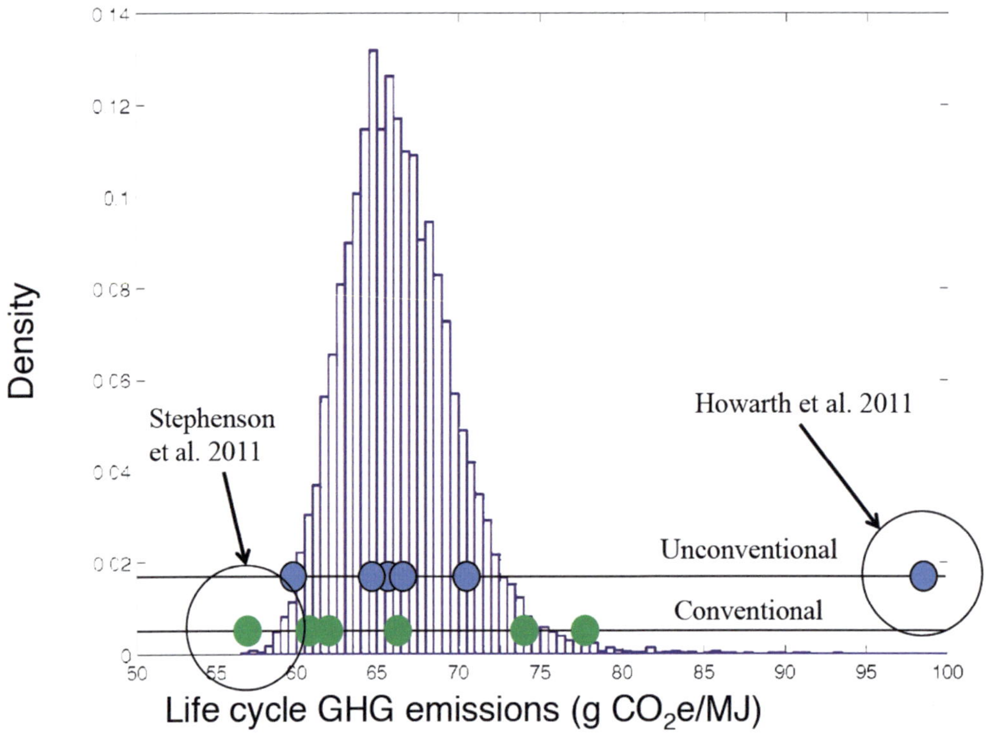

FIGURE 4.2 Probability distribution (histogram) representing the uncertainty and variability in greenhouse gas emissions from activities in the life cycle of domestic natural gas. SOURCE: Adapted with permission from Venkatesh et al. (2011b). Copyright 2011 American Chemical Society. Superimposed are estimates from studies of life-cycle emissions from unconventional and conventional hydrocarbon production (blue and green dots, respectively). SOURCE: Paulina Jaramillo, Carnegie Mellon University.

a rate of 1.25 percent per year would affect temperature, assuming that a percentage of methane leaks into the atmosphere. The study found that the switch to natural gas would produce short-term (decades) warming because methane leakage warms the atmosphere and aerosols produced by coal combustion cool the atmosphere.

Jaramillo concluded that the life-cycle greenhouse gas emissions from shale gas can be higher than those from conventional gas. However, she noted the need for additional impact assessments and climate modeling to assess the impact of shale gas on climate and for displacement analysis to assess the effect of changing fuel sources on climate. She also thought that a proper regulatory framework is needed to manage all environmental impacts.

Questions. A workshop participant noted that the maximum leakage rate (7.7 percent)[2] estimated in a National Oceanic and Atmospheric Administration study (Pétron et al., 2012) is controversial. In an analysis of the work, Levi (2012) estimated that leakage rates are closer to 1 to 2 percent. Greg Frost, a coauthor of the Pétron et al. (2012) article, said that the Levi (2012) paper correctly pointed out the unreliability of inventory information used to interpret the atmospheric measurements in terms of emissions. However, alternative methods produce the same results: a leakage

[2] Pétron et al. (2012) report natural gas losses ranging from 2.3 to 7.7 percent, with an expected value of 4.4 percent.

rate of about 4 percent in the Denver-Julesburg Fossil Fuel Basin in Colorado and about 9 percent in Utah. Jaramillo added that published studies suggest an overall leakage rate of about 3 percent, although rates may be higher in particular fields.

Another participant commented that two ExxonMobil researchers recently evaluated life-cycle emissions using data from Marcellus well sites, compressor stations, and other measurements (see Laurenzi and Jersey, 2013).

WORKING GROUP REPORTS

Each working group was asked to consider the following themes:

- Potential effects on landscapes, including soil and living organisms, and other environmental systems—Connections between hydraulic fracturing and other production technologies and processes on environmental systems, including scientific data and methods in assessing impacts;
- Technical and engineering processes—Methods for limiting and mitigating the impacts of developing and producing hydrocarbons from unconventional resources on landscapes, ecosystems, and climate; and
- Research priorities—Scientific and engineering research needed to narrow or characterize uncertainties.

A wide range of issues related to these themes were raised by the working groups. Issues that were raised by multiple working groups or that were raised for the first time in the workshop are summarized below. The complete working group reports appear in Appendix D.

Research Priorities and Issues

Working groups identified a wide range of environmental issues for further study. These included understanding impacts at different temporal and spatial scales, from individual species (thresholds and tipping points) to landscapes (erosion and sedimentation, spills, topographic alteration, road use, and habitat disruption) to the atmosphere (toxins and particulate matter) and climate (greenhouse gas emissions over the life cycle of the well). Some of the working groups discussed the need to compare or untangle the environmental impacts from legacy conventional wells and modern shale gas wells.

The long timescales required to study the environmental impacts of shale gas development pose problems for research (e.g., difficulties of obtaining funding for monitoring). Some working groups thought that industry might be willing to provide long-term funding for applied research and the development of low-cost monitoring networks and large data management systems. However, industry practices change much faster than research results are generated, so it is a challenge to keep research studies and partnerships relevant to industry.

Baseline and monitoring data were key issues of the working groups. Several working groups stressed the need for standardized data collection and sampling methods. Careful thought about the problem could help target data collection to the right parameters, areas, and timescales. Some working groups suggested involving citizens, extension services, or county conservation agents in data collection or monitoring. For example, the development of sensors that are inexpensive and easy to operate, such as those on weather stations operated by volunteers,[3] could significantly

[3] More than 11,000 volunteers collect weather observations as part of the National Weather Service Cooperative Observer Program. See http://www.nws.noaa.gov/om/coop/what-is-coop.html.

expand data collection. Better access to government and industry data would also increase the pool of data available for research.

Public Perception and Education

Some of the working groups noted that citizens make decisions and take positions on shale gas production based on potential impacts to people (e.g., increased traffic, light, noise, and odors from generators and compression stations) or to ecosystems (e.g., habitat fragmentation or loss). Consequently, it is important that scientists communicate clearly what is known about sociological and ecological impacts. Extension services may also be able to play a role in public communication and education.

One working group pointed out that research alone is unlikely to change public perceptions. Good management could reduce the sociological impacts of light, noise, and traffic and also improve public perceptions. Proceeding conservatively with development until potentially large ecological impacts are better understood may also be a useful strategy. Some individuals said that the industry already takes ecological impacts into consideration when planning development.

Technical and Engineering Issues

The working groups identified a number of technical and engineering issues that need further attention, including placing wells, pipelines, and facilities in locations that minimize environmental impacts while optimizing hydrocarbon production; and developing techniques and strategies to reduce the effect of noise and light pollution on ecosystems and communities. Although one working group agreed that measures to reduce fugitive emissions are needed, it disagreed on where mitigation efforts should be focused (e.g., leaky pipes in cities may have higher fugitive emissions than shale gas production). One working group noted the importance of a responsible operation philosophy and best management practices.

5

Final Thoughts

The workshop concluded with final thoughts offered by the planning committee and workshop participants. George Hornberger, planning committee chair, said that, despite uncertainties, it appears that there are substantial unconventional hydrocarbon reserves and a large potential for economic development. A common theme was the need for baseline data and ongoing data collection for monitoring environmental impacts. In addition, communication is important at all levels because public perceptions play a substantial role in what can be accomplished. Addressing impacts in ways that are broadly acceptable can be facilitated by communication and collaboration within and across university, government, and industry sectors. Finally, understanding the costs (e.g., environmental impacts) and benefits (e.g., economic gain) over the life cycle of shale gas development is critical for informing decisions.

Kate Baker agreed that scientists and engineers have difficulty communicating among themselves, and that it is even harder to talk to (and listen to) people outside the field. There are compelling needs for baseline data. Resources (money, people, time) are insufficient to measure everything, so some thought has to go into collecting the right measurements in the right places at the right time and with the right resolution and accuracy to address key issues or problems. Finally, the workshop discussions reminded her that all shales, ecosystems, communities, and political systems are different. However, some lessons learned will translate to other places and we as individuals should be mindful of sharing them across our different communities and jurisdictional boundaries.

Michael Hohn focused on the geology and engineering issues that opened the workshop. Although the community has learned a great deal over the past few decades, fundamental work remains to be done. Examples include better well placement, improved predictions of well performance, better characterization of the internal structure of shales, and accurate assessments of the resource at spatial scales ranging from an individual well to the entire nation. A better understanding of what is going on would help make the process more efficient, which would also reduce adverse impacts to the environment.

Susan Brantley said that experience has shown that if the public has a concern, scientists have to devote resources to it, even if they do not agree with the concern. The speed of shale gas development in Pennsylvania and some of the mistakes that were made may have led to pushback elsewhere. For example, hydraulic fracturing is now illegal in France. Listening to what the public thinks, responding to their concerns, and making good environmental decisions is probably a good long-term strategy. She concluded by describing an effort (Shale Network)[1] aimed at collecting water quality and water quantity data from university groups, federal and state government entities, and others in areas where natural gas is being extracted. The data are being made available online so they can be readily found and analyzed.

Hornberger invited comments from workshop participants, and William Kappel (U.S. Geological Survey [USGS]) described a research and monitoring strategy for managing unconventional hydrocarbon development in the Appalachian Basin. This integrated science plan is being developed by the USGS and other federal agencies. Following federal agency review, input will be sought from academia, industry, and state regulators to determine (1) whether an integrated approach will address some of the issues raised in the workshop and (2) whether data are available to start analyzing these issues. The objective is to gather data and place them in a usable format so that people can understand how they are likely to be affected by shale gas production.

The final remarks were made by Fred King (West Virginia University), who saw several sets of challenges and opportunities. First, despite public fears and extreme comments on both sides, it is up to the science and engineering community to work together and better educate the public. Increasing energy literacy will help the public understand the issues and make rational decisions. Social science and policy could be used to help people understand the regulations and to make those regulations more uniform across geographic areas. Another opportunity is filling gaps in knowledge through both basic research and applied research (e.g., best practices, best ways to exploit the resource). Finally, funding is a challenge because federal resources are insufficient for all the research or data collection that is needed, and a strong business case has to be made to convince industry to contribute research funding. Overall, he saw a tremendous opportunity for a partnership between industry, academia, policy makers, and the public to decide how to handle this natural resource in a way that benefits the region as a whole.

[1] See http://www.shalenetwork.org/.

Bibliography

The following references to published articles were cited in the workshop presentations or shared by participants during the workshop.

Adams, M.B., P.J. Edwards, W.M. Ford, J.B. Johnson, T.M. Schuler, M. Thomas-Van Gundy, and F. Wood. 2011. Effects of Development of a Natural Gas Well and Associated Pipeline on the Natural and Scientific Resources of the Fernow Experimental Forest. Newtown Square, PA: U.S. Department of Agriculture, 24 pp.

API (American Petroleum Institute). 2009. HF1, Hydraulic Fracturing Operations—Well Construction and Integrity Guidelines, 1st Ed. Available at http://www.api.org/policy-and-issues/policy-items/hf/api_hf1_hydraulic_fracturing_operations.

API. 2010. HF2, Water Management Associated with Hydraulic Fracturing, 1st Ed. Available at http://www.api.org/oil-and-natural-gas-overview/exploration-and-production/hydraulic-fracturing/api_hf2_water_management.

API. 2011. HF3, Practices for Mitigating Surface Impacts Associated with Hydraulic Fracturing, 1st Ed. Available at http://www.api.org/policy-and-issues/policy-items/hf/api_hf3_practices_for_mitigating_surface.

Baker, M.E., and R.S. King. 2010. A new method for detecting and interpreting biodiversity and ecological community thresholds. Methods in Ecology and Evolution 1:25-37.

Baldassare, F.J., M. McCaffrey, and J. Harper. 2013. A geochemical context for stray gas investigations in the N. Appalachian Basin: Implications of analyses of natural gases from Neogene-through-Devonian-Age strata. American Association of Petroleum Geologists Bulletin. doi:10.1306/06111312178.

Barbot, E., N.S. Vidic, K.B. Gregory, and R.D. Vidic. 2013. Spatial and temporal correlation of water quality parameters of produced waters from Devonian-age shale following hydraulic fracturing. Environmental Science & Technology 47:2562-2569.

Barbour, M.T., J. Gerritsen, B.D. Snyder, and J.B. Stribling. 1999. Rapid Bioassessment Protocols for Use in Streams and Wadeable Rivers: Periphyton, Benthic Macroinvertebrates and Fish, 2nd Ed. EPA 841-B-99-002. Washington, DC: Environmental Protection Agency.

BC Oil and Gas Commission. 2012. Investigation of Observed Seismicity in the Horn River Basin. British Columbia, 29 pp. Available at http://www.bcogc.ca/investigation-observed-seismicity-horn-river-basin.

Bearer, S., E. Nicholas, T. Gagnolet, M. DePhilip, T. Moberg, and N. Johnson. 2012. Evaluating the scientific support of conservation best management practices for shale gas extraction in the Appalachian Basin. Environmental Practice 14:308-319.

Blauch, M.E., R.R. Myers, T.R. Moore, and B.A. Lipinski. 2009. Marcellus Shale post-frac flowback waters—Where is all the salt coming from and what are the implications? In Proceedings of the Society of Petroleum Engineers Eastern Regional Meeting, September 23-25. SPE 125740, pp. 1-20.

Boswell, R., and T.S. Collett. 2011. Current perspectives on gas hydrate resources. Energy & Environmental Science 4:1206-1215.

Breen, K.J., C.G. Angelo, R.W. Masters, and A.C. Sedam. 1985. Chemical and isotopic characteristics of brines from three oil- and gas-producing sandstones on eastern Ohio, with applications to the geochemical tracing of brine sources. U.S. Geological Survey Water Resources Investigations Report 84-4314, 58 pp.

Capo, R.C., B.W. Stewart, E.L. Rowan, C.A. Kolesar, A.J. Wall, E.C. Chapman, R.W. Hammack, and K.T. Schroeder. In press. The strontium isotopic evolution of Marcellus Formation produced waters, southwestern Pennsylvania. International Journal of Coal Geology.

Chapman, E.C., R.C. Capo, B.W. Stewart, C.S. Kirby, R.W. Hammack, K.T. Schroeder, and H.M. Edenborn. 2012. Geochemical and strontium isotope characterization of produced waters from Marcellus Shale natural gas extraction. Environmental Science & Technology 46:3545-3553.

Chapman, E.C., R.C. Capo, B.W. Stewart, R.S. Hedin, T.J. Weaver, and H.M. Edenborn. 2013. Strontium isotope quantification of siderite, brine and acid mine drainage contributions to high-TDS abandoned gas well discharges in the Appalachian Plateau. Applied Geochemistry 31:109-118.

Chutter, F.M. 1969. The effects of silt and sand on the invertebrate fauna of streams and rivers. Hydrobiologia 34:57-76.

Clark, C.E., and J.A. Veil. 2009. Produced Water Volumes and Management Practices in the United States. ANL/EVS/R-09/1. Argonne, IL: U.S. Department of Energy, 60 pp.

Coleman, J.L., R.C. Milici, T.A. Cook, R.R. Charpentier, M. Kirschbaum, T.R. Klett, R.M. Pollastro, and C.J. Schenk. 2011. Assessment of Undiscovered Oil and Gas Resources of the Devonian Marcellus Shale of the Appalachian Basin Province. U.S. Geological Survey Fact Sheet 2011-3092, 2 pp. Available at http://pubs.usgs.gov/fs/2011/3092/.

Cordone, A.J., and D.W. Kelley. 1961. The influences of inorganic sediment on the aquatic life of streams. California Fish and Game 47:189-228.

de Pater, C.J., and S. Baisch. 2011. Geomechanical Study of Bowland Shale Seismicity: Synthesis Report. Cuadrilla Resources Ltd, 57 pp.

EIA (U.S. Energy Information Administration). 2013a. Annual Energy Outlook 2013, with Projections to 2040. DOE/EIA-0383. Washington, DC: U.S. Department of Energy, 233 pp.

EIA. 2013b. Technically Recoverable Shale Oil and Shale Gas Resources: An Assessment of 137 Shale Formations in 41 Countries Outside the United States. Washington, DC: U.S. Department of Energy. Available at http://www.eia.gov/analysis/studies/worldshalegas/pdf/fullreport.pdf?zscb=21500862.

Eckert, C., E. Ober, and S. McCallum. 2013. Sequence stratigraphic framework approach for the identification and mapping of Upper Devonian siltstones in northern West Virginia, Appalachian Basin. Presented at AAPG Annual Convention, Pittsburgh, PA, May 19-22.

Entrekin, S., M. Evans-White, B. Johnson, and E. Hagenbuch. 2011. Rapid expansion of natural gas development poses a threat to surface waters. Frontiers in Ecology and the Environment 9:503-511.

Erenpreiss, M.S., L.H. Wickstrom, C.J. Perry, R.A. Riley, D.R. Martin, and others. 2011. Regional organic-thickness map of the Marcellus Shale with additional organic-rich shales beds in the Hamilton Group included for New York, Pennsylvania, and West Virginia. Ohio Department of Natural Resources, Division of Geological Survey. Available at http://www.dnr.state.oh.us/LinkClick.aspx?fileticket=5TdkdpILE4c%3D&tabid=23014.

Francis, C.D., and J.R. Barber. 2013. A framework for understanding noise impacts on wildlife: An urgent conservation priority. Frontiers in Ecology and the Environment 11:305-313.

Frohlich, C. 2012. Two-year survey comparing earthquake activity and injection well locations in the Barnett Shale, Texas. Proceedings of the National Academy of Sciences 109:13934-13938.

Frohlich, C., and M. Brunt. 2013. Two-year survey of earthquakes and injection/production wells in the Eagle Ford Shale, Texas, prior to the MW4.8 20 October 2011 earthquake. Earth and Planetary Science Letters 379:53-63.

Frohlich, C., C. Hayward, B. Stump, and E. Potter. 2011. The Dallas-Fort Worth earthquake sequence: October 2008 through May 2009. Bulletin of the Seismological Society of America 101:327-340.

Frohlich, C., J. Glidewell, and M. Brunt. 2012. Location and felt reports for the 25 April 2010 mbLG3.9 earthquake near Alice, Texas: Was it induced by petroleum production? Bulletin of the Seismological Society of America 102:457-466.

Gan, W., and C. Frohlich. 2013. Gas injection may have triggered earthquakes in the Cogdell oil field, Texas. Proceedings of the National Academy of Sciences. doi:10.1073/pnas.1311316110.

Gaston, K.J., T.W. Davies, J. Bennie, and J. Hopkins. 2012. Reducing the ecological consequences of night-time light pollution: Options and developments. Journal of Applied Ecology 49:1256-1266.

Gillen, J.L., and E. Kiviat. 2012. Hydraulic fracturing threats to species with restricted geographic ranges in the Eastern United States. Environmental Practice 14:320-331.

Gregory, K.B., R.D. Vidic, and D.A. Dzombak. 2011. Water management challenges associated with the production of shale gas by hydraulic fracturing. Elements 7:181-186.

Haluszczak, L.O., A.W. Rose, and L.R. Kump. 2013. Geochemical evaluation of flowback brine from Marcellus gas wells in Pennsylvania, USA. Applied Geochemistry 28:55-61.

Hayes, T. 2009. Sampling and Analysis of Water Streams Associated with the Development of Marcellus Shale Gas: Final Report. Marcellus Shale Coalition, 249 pp.

Howarth, R.W., R. Santoro, and A. Ingraffea. 2011. Methane and the greenhouse-gas footprint of natural gas from shale formations. Climatic Change 106:679-690.

IOM (Institute of Medicine). 2013. Health Impact Assessment of Shale Gas Extraction: Workshop Summary. Washington, DC: The National Academies Press, 156 pp.

Jiang, M., W.M. Griffin, C. Hendrickson, P. Jaramillo, J. VanBriesen, and A. Venkatesh. 2011. Life cycle greenhouse gas emissions of Marcellus Shale gas. Environmental Research Letters 6, doi:10.1088/1748-9326/6/3/034014.

Johnson, N. 2010. Pennsylvania Energy Impacts Assessment. Report 1: Marcellus Shale Natural Gas and Wind. Harrisburg, PA: The Nature Conservancy, 47 pp. Available at http://www.nature.org/media/pa/tnc_energy_analysis.pdf.

Justinic, A., B. Stump, C. Hayward, and C. Frohlich. 2013. Analysis of the Cleburne, Texas, earthquake sequence from June 2009 to June 2010. Bulletin of the Seismological Society of America 103:3083-3093.

Kappel, W.M. 2013. Dissolved methane in groundwater, Upper Delaware River Basin, Pennsylvania and New York, 2007–12. U.S. Geological Survey, Open-File Report 2013-1167, 6 pp. Available at http://pubs.usgs.gov/of/2013/1167/.

Kappel, W.M., J.H. Williams, and Z. Szabo. 2013. Water resources and shale gas/oil production in the Appalachian Basin—Critical issues and evolving developments. U.S. Geological Survey, Open-File Report 2013-1137, 12 pp. Available at http://pubs.usgs.gov/of/2013/1137.

Karr, J.R., and D.R. Dudley. 1981. Ecological perspective on water quality goals. Environmental Management 5:55-68.

Kenny, J.F., N.L. Barber, S.S. Hutson, K.S. Linsey, J.K. Lovelace, and M.A. Maupin. 2009. Estimated Use of Water in the United States in 2005. U.S. Geological Survey Circular 1344, 52 pp.

King, R.S., M.E. Baker, D.F. Whigham, D.E. Weller, T.E. Jordan, P.F. Kazyak, and M.K. Hurd. 2005. Spatial considerations for linking watershed land cover to ecological indicators in streams. Ecological Applications 15:137-153.

Kirschbaum, M.A., C.J. Schenk, T.A. Cook, R.T. Ryder, R.R. Charpentier, T.R. Klett, S.B. Gaswirth, M.E. Tennyson, and K. J. Whidden. 2012. Assessment of Undiscovered Oil and Gas Resources of the Ordovician Utica Shale of the Appalachian Basin Province. U.S. Geological Survey Fact Sheet 2012-3116, 6 pp. Available at http://pubs.usgs.gov/fs/2012/3116/.

Krauss, R.F. 2013. Addressing well and field infrastructure siting challenges in the wetlands and streams of the Haynesville, Marcellus, Utica and Eagle Ford shale plays. Proceedings of the Society of Petroleum Engineers Americas E&P Health, Safety, Security and Environmental Conference, Galveston, TX, March 19-20. SPE 163804.

Kuntz, R., J. Ashbaugh, B. Poedjono, J. Zabaldano, I. Shevchenko, and C. Jamerson. 2011. Pad design key for Marcellus drilling. The American Oil and Gas Reporter, April. Available at http://www.slb.com/~/media/Files/drilling/industry_articles/201104_aogr_drilling_april.pdf.

Ladlee, J., and J. Jacquet. 2011. The implications of multi-well pads in the Marcellus Shale. Research & Policy Brief Series 43. Available at http://cardi.cornell.edu/cals/devsoc/outreach/cardi/publications/.

Laughrey, C.D., and F.J. Baldassare. 1998. Geochemistry and origin of some natural gases in the Plateau Province, Central Appalachian Basin, Pennsylvania and Ohio. American Association of Petroleum Geologists Bulletin 82:317-335.

Laurenzi, I.J., and G.R. Jersey. 2013. Life cycle greenhouse gas emissions and freshwater consumption of Marcellus Shale gas. Environmental Science & Technology 47:4896-4903.

Leventhal, J.S. 1981. Pyrolysis-gas chromatography to characterize organic matter and its relationship to uranium in Appalachian Devonian shales. Geochimica et Cosmochimica Acta 45:883-889.

Leventhal, J.S., J.G. Crock, and M.J. Malcolm. 1981. Geochemistry of trace elements and uranium in Devonian shales of the Appalachian Basin. U.S. Geological Survey Open-File Report 81-778, 73 pp.

Levi, M.A. 2012. Comment on "Hydrocarbon emissions characterization in the Colorado Front Range: A pilot study" by Gabrielle Pétron et al. Journal of Geophysical Research 117, doi:10.1029/2012JD017686.

Maloney, K.O., M. Schmid, and D.E. Weller. 2012. Applying additive modelling and gradient boosting to assess the effects of watershed and reach characteristics on riverine assemblages. Methods in Ecology and Evolution 3:116-128.

McBroom, M., T. Thomas, and Y. Zhang. 2012. Soil erosion and surface water quality impacts of natural gas development in East Texas, USA. Water 4:944-958.

Milici, R.C., R.T. Ryder, C.S. Swezey, R.R. Charpentier, T.A. Cook, R.A. Crovelli, T.R. Klett, R.M. Pollastro, and C.J. Schenk. 2002. Assessment of Undiscovered Oil and Gas Resources of the Appalachian Basin Province, 2002. U.S. Geological Survey Fact Sheet 009-03. Available at http://pubs.usgs.gov/fs/fs-009-03/.

Molofsky, L.J., J.A. Connor, A.S. Wylie, T. Wagner, and S.K. Farhat. 2013. Evaluation of methane sources in groundwater in Northeastern Pennsylvania. Groundwater 51:333-349.

Myers, T. 2012. Potential contaminant pathways from hydraulically fractured shale to aquifers. Groundwater 50:872-882.

Newcombe, C.P., and D.D. MacDonald. 1991. Effects of suspended sediments on aquatic ecosystems. North American Journal of Fisheries Management 11:72-82.

NETL (National Energy Technology Laboratory). 2009. Modern Shale Gas Development in the United States: A Primer. Prepared by the Ground Water Protection Council and ALL Consulting, 96 pp. Available at http://www.netl.doe.gov/technologies/oil-gas/publications/EPreports/Shale_Gas_Primer_2009.pdf.

NETL. 2011. Life Cycle Greenhouse Gas Inventory of Natural Gas Extraction, Delivery and Electricity Production. DOE/NETL-2011/1522. Available at http://www.netl.doe.gov/energy-analyses/pubs/NG-GHG-LCI.pdf.

NRC (National Research Council). 2013. Induced Seismic Potential in Energy Technologies. Washington, DC: The National Academies Press, 262 pp.

NRC. In preparation. Risk Management and Governance Issues in Shale Gas Development: A Summary of Two Workshops. Washington, DC: The National Academies Press.

ODNR (Ohio Department of Natural Resources). 2012. Preliminary Report on the Northstar 1 Class II Injection Well and the Seismic Events in the Youngstown, Ohio, Area. 23 pp. Available at http://ohiodnr.com/downloads/northstar/UICreport.pdf.

Osborn, S.G., and J.C. Macintosh. 2010. Chemical and isotopic tracers of the contribution of microbial gas in Devonian organic-rich shales and reservoir sandstones, northern Appalachian Basin. Applied Geochemistry 25:456-471.

Osborn, S.G., A. Vengosh, N.R. Warner, and R.B. Jackson. 2011. Methane contamination of drinking water accompanying gas-well drilling and hydraulic fracturing. Proceedings of the National Academy of Sciences 108:8172-8176.

Papoulias, D.M., and A.L. Velasco. 2013. Histopathological analysis of fish from Acorn Fork Creek, Kentucky, exposed to hydraulic fracturing fluid releases. Southeastern Naturalist 12(Special Issue 4):92-111.

Patrick, R. 1951. A proposed biological measure of stream conditions. Proceedings of the International Association of Theoretical and Applied Limnology 11:299-307.

Paul, M.J., and J.L. Meyer. 2001. Streams in the urban landscape. Annual Review of Ecology and Systematics 32:333-365.

Pétron, G., G. Frost, B.R. Miller, A.I. Hirsch, S.A. Montzka, A. Karion, M. Trainer, C. Sweeney, A.E. Andrews, and L. Miller. 2012. Hydrocarbon emissions characterization in the Colorado Front Range: A pilot study. Journal of Geophysical Research 117:D04304, doi:10.1029/2011JD016360.

Phan, T.T., R.C. Capo, B.W. Stewart, S. Sharma, and J. Toro. 2013. Uranium partitioning and isotope composition in shales of the Middle Devonian Marcellus Formation. Mineralogical Magazine. doi:10.1180/minmag.2013.077.5.16.

Poff, N.L., B.D. Richter, A.H. Arthington, S.E. Bunn, R.J. Naiman, E. Kendy, M. Acreman, C. Apse, B.P. Bledsoe, M.C. Freeman, J. Henriksen, R.B. Jacobson, J.G. Kennen, D.M. Merritt, J.H. O'Keeffe, J.D. Olden, K. Rogers, R.E. Tharme, and A. Warner. 2010. The ecological limits of hydrologic alteration (ELOHA): A new framework for developing regional environmental flow standards. Freshwater Biology 55:147-170.

Reid, L.M., and T. Dunne. 1984. Sediment production from forest road surfaces. Water Resources Research 20:1753-1761.

Révész, K.M., K.J. Breen, A.J. Baldassare, and R.C. Burruss. 2010. Carbon and hydrogen isotopic evidence for the origin of combustible gases in water-supply wells in north-central Pennsylvania. Applied Geochemistry 25:1845-1859.

Ritchie, J.C. 1972. Sediment, fish, and fish habitat. Journal of Soil and Water Conservation 27:124-125.

Rowan, E.L., M.A. Engle, C.S. Kirby, and T.F. Kraemer. 2011. Radium content of oil- and gas-field produced waters in the northern Appalachian Basin (USA): Summary and discussion of data. U.S. Geological Survey Scientific Investigations Report 2011-5135, 31 pp. Available at http://pubs.usgs.gov/sir/2011/5135/.

Royal Academy of Engineering. 2012. Shale Gas Extraction in the UK: A Review of Hydraulic Fracturing. London, UK: The Royal Society, 75 pp. Available at http://royalsociety.org/uploadedFiles/Royal_Society_Content/policy/projects/shale-gas/2012-06-28-Shale-gas.pdf.

Ryan, P.A. 1991. Environmental effects of sediment on New Zealand streams. New Zealand Journal of Marine and Freshwater Research 25:207-221.

Saiers, J.E., and E. Barth. 2012. Comment on "Potential contaminant pathways from hydraulically fractured shale aquifers." Groundwater 50:826-828.

Schlegel, M.E., A. Zhou, J.C. McIntosh, C.J. Ballentine, and M.A. Person. 2011. Constraining the timing of microbial methane generation in an organic-rich shale using noble gases, Illinois basin, USA. Chemical Geology 267:27-40.

Smith, D.R., C.D. Snyder, N.P. Hitt, J.A. Young, and S.P. Faulkner. 2012. Shale gas development and brook trout: Scaling best management practices to anticipate cumulative effects. Environmental Practice 14:366-381.

Soeder, D.J., B.W. Stewart, N. Pekney, L. Hopkinson, R. Dilmore, B.G. Kutchko, S. Sharma, C. Carter, J.A. Hakala, and R.C. Capo. In press. U.S. DOE methods for assessing the environmental risks of shale gas development. International Journal of Coal Geology.

STAC (Scientific and Technical Advisory Committee, Chesapeake Bay Program). 2013. Exploring the Environmental Effects of Shale Gas Development in the Chesapeake Bay Watershed. Publication 13-01. Edgewater, MD: STAC, 30 pp.

States, S., G. Cyprych, M. Stoner, F. Wydra, J. Kuchta, J. Monnell, and L. Casson. 2013. Brominated THMs in drinking water: A possible link to Marcellus Shale and other wastewaters. Journal – American Water Works Association 105:E432-E448.

Stephenson, T., and J. Valle. 2011. Modeling the relative GHG emissions of conventional and shale gas production. Environmental Science & Technology 45:10,757-10,764.

USGS (U.S. Geological Survey). 2002. Natural gas production in the United States. National Assessment of Oil and Gas Fact Sheet FS-113-02. Available at http://pubs.usgs.gov/fs/fs-0113-01.pdf.

Veil, J. 2012. A White Paper Summarizing the Stray Gas Incidence & Response Forum, Cleveland, Ohio, July 24-26, 2012, 48 pp. Available at http://www.veilenvironmental.com/publications/pw/stray_gas_white_paper.pdf.

Veil, J. 2013. A White Paper Summarizing a Special Session on Induced Seismicity. Ground Water Research & Education Foundation Spotlight Series, 38 pp. Available at http://www.veilenvironmental.com/publications/pw/white_paper_on_induced_seismicity.pdf.

Venkatesh, A., P. Jaramillo, W.M. Griffin, and H.S. Matthews. 2011a. Uncertainty analysis of life cycle greenhouse gas emissions from petroleum-based fuels and impacts on low carbon fuel policies. Environmental Science & Technology 45:125-131.

Venkatesh, A., P. Jaramillo, W.M. Griffin, and H.S. Matthews. 2011b. Uncertainty in life cycle greenhouse gas emissions from United States natural gas end-uses and its effects on policy. Environmental Science & Technology 45:8182-8189.

Venkatesh, A., P. Jaramillo, W.M. Griffin, and H.S. Matthews. 2012. Implications of changing natural gas prices in the United States electricity sector for SO_2, NO_X and life cycle GHG emissions. Environmental Research Letters 7:034018, doi:10.1088/1748-9326/7/3/034018.

Vidic, R.D., S.L. Brantley, J.M. Vandenbossche, D. Yoxtheimer, and J.D. Abad. 2013. Impact of shale gas development on regional water quality. Science 340, doi:10.1126/science.1235009.

Wachal, D.J., K.E. Banks, P.F. Hudak, and R.D. Harmel. 2009. Modeling erosion and sediment control practices with RUSLE 2.0: A management approach for natural gas well sites in Denton County, TX, USA. Environmental Geology 56:1615-1627.

Wall, A.J., R.C. Capo, B.W. Stewart, T.T. Phan, J.C. Jain, J.A. Hakala, and G.D. Guthrie. 2013. High throughput method for Sr extraction from variable matrix waters and 87Sr/86Sr isotope analysis by MC-ICP-MS. Journal of Analytical Atomic Spectrometry 28:1338-1344.

Wang, G., and T.R. Carr. 2012. Methodology of organic-rich shale lithofacies identification and prediction: A case study from Marcellus Shale in the Appalachian Basin. Computers & Geosciences 49:151-163.

Wang, L., and J. Lyons. 2003. Fish and benthic macroinvertebrate assemblages as indicators of stream degradation in urbanizing watersheds. Pp. 227-249 in Biological Response Signatures: Indicator Patterns Using Aquatic Communities, T.P. Simon, ed. New York: CRC Press.

Wang, L., J. Lyons, P. Kanehl, and R. Gatti. 1997. Influences of watershed land use on habitat quality and biotic integrity in Wisconsin streams. Fisheries 22:6-12.

Warner, N.R., R.B. Jackson, T.H. Darrah, S.G. Osborn, A. Down, K. Zhao, A. White, and A. Vengosh. 2012. Geochemical evidence for possible natural migration of Marcellus Formation brine to shallow aquifers in Pennsylvania. Proceedings of the National Academy of Sciences 109:11961-11966.

Waters, T.F. 1995. Sediment in Streams: Sources, Biological Effects, and Control. American Fisheries Society Monograph 7, Bethesda, MD.

Weltman-Fahs, M., and J.M. Taylor. 2013. Hydraulic fracturing and brook trout habitat in the Marcellus Shale region: Potential impacts and research needs. Fisheries 38:4-15.

Wigley, T.M.L. 2011. Coal to gas: The influence of methane leakage. Climatic Change 108:601-608.

Williams, H., D. Havens, K. Banks, and D. Wachal. 2008. Field-based monitoring of sediment runoff from natural gas well sites in Denton County, Texas, USA. Environmental Geology 55:1463-1471.

Wilson, J., and J. Vanbriesen. 2013. Source water changes and energy extraction activities in the Monongahela River, 2009–2012. Environmental Science & Technology 47:12575-12582.

Wiseman, H. 2012. Fracturing regulation applied. Duke Environmental Law & Policy Forum. 22:361-384.

Appendix A

JOHN D. ROCKEFELLER IV
WEST VIRGINIA

United States Senate
WASHINGTON, DC 20510–4802

STATE OFFICE:
405 CAPITOL STREET, SUITE 508
CHARLESTON, WV 25301
(304) 347–5372
FAX: (304) 347–5371

NORTHERN SATELLITE OFFICE:
118 ADAMS STREET, SUITE 301
FAIRMONT, WV 26554
(304) 367–0122
FAX: (304) 367–0822

SOUTHERN SATELLITE OFFICE:
220 NORTH KANAWHA STREET, SUITE 1
BECKLEY, WV 25801
(304) 253–9704
FAX: (304) 253–2578

EASTERN REGIONAL OFFICE:
217 WEST KING STREET, SUITE 307
MARTINSBURG, WV 25401
(304) 262–9285
FAX: (304) 262–9288

September 9, 2013

Dear Friends,

Although I cannot be with you today, I want to thank the National Research Council for extending an invitation to join all of you today for this important workshop on the development of shale gas in the Appalachian Basin.

The existence of shale gas may not be new, but the technology to access it economically is – and I'm proud to see the development of this resource in West Virginia and how it is already making a significant footprint here in the State.

While this unconventional natural gas offers great benefits, as I've said many times before, it also raises very real challenges that we can and must work together to address. The development of shale gas does not need to come at the expense of environmental protections, nor should we sacrifice our beautiful lands for this resource. West Virginians deserve that from us and our environmental future depends on it. If we are to use the safest, most careful and responsible extraction methods, that are balanced with developers' needs and our communities' concerns, I am confident that West Virginia will continue to enjoy much success from the development of shale gas.

Underlying these efforts, must be a concerted effort to maintain the public's confidence in the extraction process and the safety of natural gas operations. I'm pleased West Virginia is ahead of the curve here -- the state has already taken the initial steps to update our laws and regulations governing shale gas development. The changes effectively address potential health and safety concerns while creating a more stable environment that can cultivate growth in this industry.

These are just some of the many important issues around shale gas development and production that I'm sure will be address today. I wish you all the best for a very productive workshop and I look forward to reviewing the National Research Council's full report.

Sincerely,

John D. Rockefeller IV

http://rockefeller.senate.gov

Appendix B

Workshop Agenda

Erickson Alumni Center
West Virginia University
Morgantown, West Virginia
September 9-10, 2013

Monday, September 9

8:00 Registration

8:30 Welcome — *James P. Clements, President, West Virginia University*

8:40 Overview of the workshop — *George Hornberger, Chair, Vanderbilt University*

9:00 **Session 1: Geology, hydrocarbon resources and their development, and induced seismicity**

9:00 Geology, resources, and potential activity levels
Ray Boswell, National Energy Technology Laboratory

9:20 Engineering and technology for developing unconventional hydrocarbon resources: Current and prospective methods for exploration and production
Joseph Frantz, Jr., Range Resources Corporation

9:40 Earthquakes induced by hydrocarbon production: What Texas can tell us about Appalachia
Cliff Frohlich, University of Texas

10:00 Questions

10:20 Instructions to the working groups — *George Hornberger*

10:25 Break and divide into working groups

12:00 Working lunch

1:00 Working groups report back

Working group 1	*Kate Hadley Baker or Richard Bajura*
Working group 2	*Susan Brantley or Daniel Billman*
Working group 3	*Michael Hohn or Hannah Wiseman*
Working group 4	*Carl Kirby or Radisav Vidic*

1:30 **Session 2: Water and regulations**

1:30 Water issues relating to unconventional oil and gas production
John Veil, Veil Environmental, LLC

1:50 Identifying and assessing potential impacts of unconventional hydrocarbon production on surface and groundwater quality
Rosemary Capo, University of Pittsburgh

2:10 State shale gas regulation in the Appalachian Basin: Recent enhancements, remaining gaps, and opportunities for change
Hannah Wiseman, Florida State University

2:30 Questions

2:50 Instructions to the working groups — *George Hornberger*

3:00 Break and divide into working groups

4:45 Working groups report back

Working group 1	*Kate Hadley Baker or Paul Ziemkiewicz*
Working group 2	*Susan Brantley or Gregory Frost*
Working group 3	*Michael Hohn or Zuleima Karpyn*
Working group 4	*Carl Kirby or Paulina Jaramillo*

5:15 Workshop adjourns for the day

Tuesday, September 10

8:00 Registration

8:30 Welcome and plans for the day — *George Hornberger*

8:40	**Session 3: Ecosystems, air, and climate**	
8:40	Potential impacts of unconventional hydrocarbon production on stream biota: Current and needed research *Kelly Maloney, U.S. Geological Survey*	
9:00	Assessing and minimizing ecological impacts of shale development *Michael Powelson, The Nature Conservancy*	
9:20	Assessing emissions of hydrocarbons from rural and natural gas drilling impacted areas in Pennsylvania *Jose Fuentes, Pennsylvania State University*	
9:40	Climate impacts of shale gas *Paulina Jaramillo, Carnegie Mellon University*	
10:00	Questions	
10:20	Instructions to the working groups	*George Hornberger*
10:25	Break and divide into working groups	
12:00	Working lunch	
1:00	Working groups report back	
	Working group 1	*Kate Hadley Baker or Mark Engle*
	Working group 2	*Susan Brantley or Daniel Billman*
	Working group 3	*Michael Hohn or Peter MacKenzie*
	Working group 4	*George Hornberger or Patrick Drohan*
1:30	Wrap-up and discussion	*All*
3:30	Workshop adjourns	

Appendix C

Workshop Participants

Brian Anderson, West Virginia University
Richard Bajura, West Virginia University
Kate Hadley Baker, BP, Retired
Daniel Billman, Billman Geologic Consultants, Inc.
Ray Boswell, National Energy Technology Laboratory
Susan Brantley, Pennsylvania State University
Margaret Brittingham, Pennsylvania State University
Robert Burruss, U.S. Geological Survey
Rosemary Capo, University of Pittsburgh
Tim Carr, West Virginia University
Martin Chapman, Virginia Polytechnic Institute and State University
Eugene Cilento, West Virginia University
Nigel Clark, West Virginia University
James P. Clements, West Virginia University
Jared Cohon, Carnegie Mellon University
Jay Cole, West Virginia University
John Craynon, Virginia Polytechnic Institute and State University
Jeffrey Daniels, Ohio State University
Ashley Douds, EQT
Patrick Drohan, Pennsylvania State University
Eric Edkin, National Research Council
Elizabeth Eide, National Research Council
Mark Engle, U.S. Geological Survey
Barbara Evans Fleischauer, West Virginia State Delegate
Joseph Frantz, Jr., Range Resources
Cliff Frohlich, University of Texas
Gregory Frost, University of Colorado and National Oceanic and Atmospheric Administration
Jose Fuentes, Pennsylvania State University

Christina Gabriel, University Energy Partnership
Marc Glass, Downstream Strategies, LLC
Kelvin Gregory, Carnegie Mellon University
Shawn Grushecky, West Virginia University
J. Alexandra Hakala, National Energy Technology Laboratory
Michael Hohn, West Virginia Geological and Economic Survey
George Hornberger, Vanderbilt University
Jeffrey Jacobs, National Research Council
Paulina Jaramillo, Carnegie Mellon University
Donald Juckett, GEO-DC at AAPG (Retired)
William Kappel, U.S. Geological Survey
Zuleima Karpyn, Pennsylvania State University
Abby Kinchy, Rensselaer Polytechnic Institute
Daniel King, Department of Energy
Fred King, West Virginia University
Carl Kirby, Bucknell University
Anne Linn, National Research Council
Peter MacKenzie, Ohio Oil and Gas Association
Kelly Maloney, U.S. Geological Survey
Kris Nygaard, ExxonMobil Production Co.
Thomas Parris, Kentucky Geological Survey
Douglas Patchen, West Virginia Geological and Economic Survey
Michael Powelson, The Nature Conservancy
Shannon Rengers, FTI Consulting
Nino Ripepi, Virginia Polytechnic Institute and State University
Nicholas Rogers, National Research Council
Gary Slagel, Steptoe & Johnson
Clement Solomon, West Virginia University
Richard Stafford, Carnegie Mellon University
Samuel Taylor, NAS Science and Engineering Ambassadors Program
James Van Nostrand, West Virginia University
John Veil, Veil Environmental, LLC
Radisav Vidic, University of Pittsburgh
Michele Wheatly, West Virginia University
Richard Wilkin, Environmental Protection Agency
Hannah Wiseman, Florida State University
Gregory Wrightstone, Geologic Consultant
Paul Ziemkiewicz, West Virginia University

WEBCAST PARTICIPANTS

John Ajak, Bureau of Land Management
Katharine Avary, consulting geologist
Joseph Baran, John T. Boyd Company
Linda Barker
Mary Beilby
Jerry Berry, eigenweg USA
Pamela Billman, Billman Geologic Consultants, Inc.
Maeve Boland, American Geosciences Institute

David Brezinski, Maryland Geological Survey
Amanda Burns, West Virginia University
Susan Chiang, Center for Environmental Health
Susan Christopherson, Cornell University
Sheila Cohen, SUNY Cortland Environmental Justice Committee
Morgan Copeland, West Virginia University
Lori Davias, Tioga County Conservation District
Bridget DiCosmo, Inside EPA
Bob Donnan
Richard Elliott, Department of Energy
Sandra Fallon, West Virginia University
Erica Folio, Department of Energy
Elizabeth Geltman, Hunter College
Ben Gilmer, Downstream Strategies
Court Gould, Sustainable Pittsburgh
Erin Haynes, University of Cincinnati
Steve Hull
Allan Jelacic, Department of Energy, retired
Richard King, John T Boyd Company
Mark Kozar, U.S. Geological Survey
Ronald Landy, Environmental Protection Agency
Richard Lattanzio, Congressional Research Service
P. Patrick Leahy, American Geosciences Institute
Anya Litvak, Pittsburgh Post-Gazette
Ann Maest, Buka Environmental
Luis Alberto Fernandez Marquez
Kurt McCoy, U.S. Geological Survey
Angela McFadden, Environmental Protection Agency
Dorothy Merritts, Franklin and Marshall College
Teresa Mills, Center for Health, Environment & Justice
Isabel Montanez, University of California, Davis
Claudia Mora, Los Alamos National Laboratory
Clayton Nichols, Department of Energy, retired
Olayinka Ogunsola, Department of Energy
Brian Penttila
Ron Pettengill, Epiphany Solar Water Systems
Michael Ratner, Congressional Research Service
Kim Schultz, The Endocrine Disruption Exchange
Dana Singer, Mid-Ohio Valley Health Department and Rural Health Alliance
Xueyan Song, West Virginia University
Gus Souki, Epiphany Solar Water Systems
Paul Stern, National Research Council
Dennis Stottlemyer, West Virginia Department of Environmental Protection
James Tichenor, Bureau of Land Management
Mary Tiemann, Congressional Research Service
Ken Ward, Jr., The Charleston Gazette

Appendix D

Working Group Reports

Following each plenary session, workshop participants divided into four working groups. All four working groups considered the same topics, although the groups often interpreted them differently or emphasized different points. The following points, edited slightly for grammar and clarification, were reported by the working group chairs or rapporteurs in the plenary sessions.

SESSION 1: GEOLOGY, RESOURCES, AND PRODUCTION

Working Group 1 Report

The group did not have a detailed discussion about the resource estimates presented in the plenary or about technology and engineering processes.

Communication and Education

- How do we get the message to the general public? What should the message be? How do we communicate under uncertainty? How do we galvanize action?
- Universities have a unique role in providing credible information on costs and risks of shale gas development, but they need to be transparent about funding sources.
- Local workforce development is important, not only in STEM (science [including geoscience], technology, engineering, and mathematics) fields, but also in integrating and multidisciplinary studies.
- Academics and government agencies need to understand hydraulic fracturing as well as industry does.
- A road map or common vision for developing shale gas (e.g., number of wells, expected revenue) is needed.

Research and Development Needs

- Leakage rates.
- Sound science and statistically relevant baseline measurements, including additional chemical constituents, over historically significant periods.
- Deep monitoring wells to look at vertical migration of fluids.
- Quantifying data in a statistically meaningful way to understand cause and effect.
- Relationship between well completion strategies and estimated ultimate recovery.
- Understanding induced seismicity and deploying denser seismic infrastructure, especially around high-rate injection wells.
- Shale reservoir characterization.
- Appraising everything in the proper scale context (e.g., number of wells), well performance, and the real recovery.

Working Group 2 Report

Geology and Hydrocarbon Resources

- Research to understand differences in geologic setting and process controls (depositional environment, wet vs. dry, structural implications) between the northeast and southwest portions of the Marcellus and Utica plays.
- How to calculate stimulated rock volume and relate it to gas in place, estimated ultimate recovery, and resources and reserves.
- Geologic controls on the productive, operational, and economic lifetimes of unconventional wells (anecdotal reports of Marcellus well plugging).
- Research on nanoporosity and nanopermeability.
- Determining the geologic factors that cause induced seismicity in a particular well.

Technical and Engineering Processes

- Better production data (monthly and daily) to better estimate resources and reserves for a well, field, or region.
- Better public understanding of hydraulic fracture treatment.
- Ways to measure and model stimulated rock volume.
- Engineering of casing and cement for unconventional wells.
- Relation of reservoir pressure to induced seismicity.
- Impact of unconventional operations on abandoned or improperly abandoned wells.

Research Priorities (Meta Issues)

- Ways to balance industry needs (immediate) with options for working with individual researchers, universities, or consortia.
- Developing a common language for academia and industry.
- Including public policy makers and educators in the discussion.
- Availability of data, including data currently in the public domain and data that could be made public (e.g., monthly industry data).

Working Group 3 Report

Induced Seismicity

- How much water will be disposed over next 50 years, and where? Reuse/recycling rates are high in the Marcellus but not elsewhere.
- What is known about the interval being injected (e.g., whether faults are present, how far an earthquake might be felt)?
- What lessons have been learned from geothermal energy.

Technology and Potential Impacts

- Producing as much gas as possible using as little water and as few wells as possible. More can be learned on issues such as rock structure at smaller scales and how that affects productivity.
- Abandoned wells, legacy issues for old and new wells, and well integrity.
- Issues surrounding constituents in flowback and produced water, including the local importance of naturally occurring radioactive species.
- Using technology in flexible ways to protect sensitive ecological areas (e.g., judicious selection of well pad and pipeline locations).

Core Message: Lack of Data

- Location of abandoned wells.
- Baseline data and uniform measurements of methane and other constituents in water.
- Ecological criteria.
- Three-dimensional seismic and core data.
- Extent to which fracturing in a formation will affect future activities.
- Ways to obtain and share data beyond individual company contacts.

Working Group 4 Report

Geology

- Resources are vast, although estimates change continuously.
- The formations are spatially diverse, which makes them difficult to fully characterize. Shale is the least researched rock. Better subsurface characterization tools are needed.
- Nonproprietary data collected by states in our region are not easily accessible.

Resource Development

- The industry is relatively new, but is developing fast through an iterative process.
- The proprietary nature of the technology affects research and development in academia and government.
- The lack of public data affects further development and erodes the public trust. Industry and regulators could decide which data should remain proprietary and what additional data could be made public.
- Regional and cumulative impacts (economic, environmental, social) are difficult to assess by a single sector.
- Regulations can affect the footprint.

- Highly interdisciplinary expertise is needed to tackle some of the issues above.
- Federal and state funding for research and development is insufficient. The public needs an honest broker to understand which reports are credible.
- Induced seismicity is poorly understood.
- Better public communication, outreach, and education are needed.

SESSION 2: WATER AND REGULATIONS

Working Group 1 Report

Best Management Practices

- Well completions to minimize surface water and shallow groundwater contamination.
- Baseline monitoring and determination of what constitutes the baseline.
- Standardization of parameters (e.g., water quality) and sampling protocols.
- Standardization of criteria for measuring contamination.
- Relieving land owners of liability from test results of domestic wells.

Wellbore Integrity

- Quantify the potential for migration of contaminants from the fracture zone into drinking water aquifers.
- Identify methods for detecting leaks.
- Develop better cements.
- Carry out hydrostatic testing prior to well completion.
- Use suitable standards for testing wellbore integrity.
- Use well testing protocols (e.g., downhole tools).

Regulatory Technology and Practice

- Site and road construction and reclamation.
- Erosion control methods, inspection, and enforcement.
- Methods for managing pit and tank bottoms and sediment.
- Process for updating state and multistate regulations and standards, which vary across state lines and may conflict with one another.
- Longitudinal studies on long-term processes and impacts to water quality, and communication of these results to the public.
- Development of an Appalachian brine and natural gas geochemistry library.
- Legal and regulatory rationale and enforcement metrics.
- Development of groundwater standards irrespective of oil and gas development.
- Development of better, more cost-effective water treatment technologies.

Working Group 2 Report

Effects on Surface Water Quality and Quantity

- Collecting baseline data and making them accessible to researchers and the public across jurisdictional boundaries.
- Engaging citizens in collecting monitoring data.

- Better understanding of naturally occurring radioactive material.
- Subsurface fluid migration—what happens to fluids that do not come out of the well?
- Long-term impacts of sedimentation and erosion.

Technical and Engineering Processes

- Best practices for casings, cements, and other aspects of well construction.
- Integrity of legacy wells and influence of current activity on these wells.
- Reducing the volume of water for hydraulic fracturing.
- High-rate, in situ treatment techniques for flowback and produced water.
- Recovering constituents from wastewater for other uses.
- Continued evolution of green practices.

Meta Issues

- Data, data, data.
- Two-way communication with the public.
- Availability of injection wells, particular in Pennsylvania.
- Multiple jurisdictions, which result in different regulations across the region and hinders data collection and sharing.

Working Group 3 Report

Facts

- The uniqueness of Appalachian Basin brings challenges and opportunities:
  - brine composition;
  - total dissolved solids issues, including the scalability of existing regulatory norms for shale gas production; and
  - turbidity.
- Rapid changes in water management practices call for adaptability of regulatory agencies.
- Need for baseline testing is critical.
- Regulations in the Appalachian region are applied differently.

Research Priorities

- Identify first-order parameters for baseline studies.
- Identify vulnerable points and process gaps in water management to prioritize where regulations would be needed.
- Determine what to do with produced water and cuttings.
- Form neutral groups (e.g., independent research, university consortia) to assess the impacts of shale gas production.
- Identify real and perceived risks and mitigation strategies.
- Provide information to help the public differentiate between perceived and actual risk.
- Develop advanced water treatment methods.
- Develop high-end desalination processes to both improve water quality and recover valuable compounds.

- Monitoring:
  - data collection to build a database to support high-impact research,
  - seismic monitoring, and
  - air quality modeling.
- Develop better predictive tools for groundwater migration and fate.

Working Group 4 Report

Potential Effects on Surface Water and Groundwater Quality and Quantity

- Resources and inspectors are insufficient to guarantee regulatory compliance. Retention and salaries are an issue.
  - Disagreement: How much regulation/taxing is too much? Is industry already ahead of regulation?
- Chemicals or materials used during preproduction or production stages (not the hydraulic fracturing stage) have appeared in surface water or groundwater. Are they a concern for water quality?
  - Disclosure of all the materials used in the industry is needed. Some of this is happening.
- How can solids, some of which contain radioactive materials, be disposed after produced water is treated?
- Although aggregate water consumption is comparatively low, where the water is withdrawn (e.g., a small stream) influences its impact.

Technical and Engineering Processes

- Maintaining the integrity of new and aging wells. What can be done with abandoned wells to make sure groundwater is not being contaminated?
  - Develop sensors to monitor the infrastructure.
  - Can/should sensors or monitoring tools be used in "shallower" vertical wells to monitor what is happening in the deeper shale wells?
- Could alternative water sources such as acid mine drainage be used?

Research Priorities

- Are there any worrisome materials used in small quantities for which not enough information is available?
- Amount of water coming out during the life of the well and how it will be disposed in the long term.
- Potential problems with radionuclides in wastewater and solid waste from shale production.
- Variability of stray methane in well water and baselines to detect methane migration.

SESSION 3: ECOSYSTEMS, AIR, AND CLIMATE

Working Group 1 Report

Monitoring for Ecosystems, Air, and Climate

- Sampling strategies—where data are collected relative to the baseline location can bias results.

• Long-term monitoring and baseline data are needed, but it is difficult to find and maintain funding. Collaboration between industry and government and university researchers could bring new sources of funding, but might require more disclosure than companies want.

• Is it possible to untangle long-term impacts from legacy wells from short-term impacts from unconventional hydrocarbon development? Is that necessary, given that a nonpristine system is being perturbed?

• Are monitoring data covering the right areas, parameters, and timescales?

Public Perception

• Every small incident is interpreted in the context of the larger stressed environment (e.g., increased traffic, light, noise, odors from generators and compression stations).

• Will more data collection really change public attitudes about development activities?

• What are the potential impacts and risks? How much change will the public accept? Is cost-benefit analysis the best way to capture these issues?

• There is a difference between public and scientific perceptions of risk.

• Citizens make decisions and take positions based on potential ecological impacts. Scientists need to be clear about what is known about these impacts.

• For important issues with very large uncertainties, it may be best to act conservatively until impacts are better understood.

R&D and Educational Opportunities

• Baseline data and monitoring:
  ○ Use extension and county conservation agents for monitoring, although they are often overwhelmed and underfunded.

• Public understanding of ecological impacts:
  ○ To what extent should public perception guide research?
  ○ Extension agents may provide a communication channel.

Working Group 2 Report

Potential Effects on Landscapes, Including Soil, Organisms, and Systems

• Sampling methodologies (when, where, how, etc.). Geologic sampling is a science.

• Further study of water use, reuse, treatment, disposal, geochemistry, and their linkages.

• Air issues, including direct toxins and particulate matter.

• Greenhouse gas emissions in the context of the life cycle of the well.

• Surface impacts, including erosion and sedimentation, surface spills, topographic alteration, road use, and habitat disruption.

Off subject: Sociological impacts of light, noise, and traffic need to be better understood. Reducing impacts may require less research and more good management, and could improve public perceptions.

Technical and Engineering Processes

• Environmental impact statements as related to shale gas in numerous areas, basins, etc. (state vs. federal review).

- How data on greenhouse gas emissions are used and put into context. Given that industry practices are constantly changing, how does research get ahead so it has an impact?
- Keeping up with numerous and changing practices in wastewater treatment, disposal, transport, reuse, injection, and subsurface transport.
- Surface and engineering geologic input on placing wells, pipelines, and facilities in locations that minimize environmental impacts while optimizing hydrocarbon production.
- Development and use of sensors to monitor fluids, air, biota, etc. on well sites and surrounding areas.

Research Priorities (Meta Issues)

- Time lines and mechanisms for obtaining funding.
- Time lines for carrying out a study and for making measurement long enough to be useful.
- Metrics for determining the social benefits and actual impacts (i.e., water demands).
- Data issues:
  - Better access to government-collected legacy and real-time data;
  - More succinct definition of the problem, which will help define what data are needed;
  - Ways to share industry data with academia and government.
- Comparisons of impacts from legacy well drilling and completion and shale gas well drilling and completion.
- Ways to keep studies and partnerships relevant when industry often changes its best practices.

Working Group 3 Report

Major Areas of Agreement

- Importance of understanding thresholds for maintaining stable development without adversely affecting ecosystems.
- Need for standardized data collection and methods.
- Need to collect baseline data (air, water, etc.).
- Challenge of addressing issues across jurisdictional boundaries.
- Need to diversify funding in an era of declining federal funding.

Major Areas of Disagreement

- Industry consideration of habitat impact and other issues. There is significant distrust of what industry is doing.
- Where to focus mitigation of fugitive emissions. The largest sources may not be drilling and production, but instead leaky pipes in cities or leaks in other parts of the system.

Areas of New Research and Development

- Relationship between drilling or production activity, seasonality, and impacts.
- Standards (constituents, data gathering).
- Thresholds and tipping points for core species.
- Development of sensors that are cheap and easy to operate by citizens (e.g., backyard weather stations).

Working Group 4 Report

Potential Effects on Landscapes

- Ecosystem fragmentation and community and social impacts (e.g., gas production can discourage farming).
- Stormwater management and the importance of documenting the acute effects of a large storm and the cumulative effects of erosion and sedimentation.
- Effect of legal challenges on where and how development occurs.

Technical and Engineering Processes

- Ways to reduce the effect of noise (e.g., from compressor stations) and light pollution (e.g., from flaring, lights on infrastructure) on ecosystems and communities.
- Responsible operation philosophy and best management practices:
    - Enforcement or regulation when behavior not in line with the regulations;
    - Stakeholder discussion of best management practice implementation to address the scale of shale gas disturbance;
    - Regulatory staffing issues (i.e., insufficient number of qualified people).

Research Priorities

- Establish a collaboration between industry and government:
    - Rapid implementation;
    - Long-term funding for applied research;
    - Low-cost monitoring solutions, large data management systems, and better use of industry gray literature.
- Identify the externalities or unintended consequences of unconventional development in the near and long terms.
- Provide public education about the true cost of energy.
- Use university extension services to increase the research-education feedback.

Appendix E

Biographical Sketches of Planning Committee Members

George Hornberger, *Chair*, is distinguished university professor at Vanderbilt University, where he also directs the Vanderbilt Institute for Energy and the Environment. He also has a shared appointment there as the Craig E. Philip Professor of Engineering and as Professor of Earth and Environmental Sciences. He previously was a professor at the University of Virginia for many years where he held the Ernest H. Ern Chair of Environmental Sciences. He has been a visiting scholar at the Australian National University, Lancaster University, Stanford University, the U.S. Geological Survey (USGS), the University of Colorado, and the University of California, Berkeley. His research is aimed at understanding complex water–energy–climate interrelationships and at how hydrological processes affect the transport of dissolved and suspended constituents through catchments and aquifers. He is an Institute for Scientific Information Highly Cited Researcher in environmental sciences and engineering, a recognition given to the top 250 individual researchers in each of 21 subject categories. Dr. Hornberger is a fellow of the American Geophysical Union (AGU), the Geological Society of America, and the Association for Women in Science. He was president of the Hydrology Section of AGU from 2006 to 2008. He has been a member of the Nuclear Waste Technical Review Board (a Presidential appointment) since April 2004. He has served on numerous boards and committees of the National Research Council, including as chair of the Commission on Geosciences, Environment, and Resources (1996-2000) and chair of the Board on Earth Sciences and Resources (2003-2009). Dr. Hornberger won the Robert E. Horton Award (Hydrology Section) from the AGU in 1993. In 1995, he received the John Wesley Powell Award from the USGS. In 1999, he was presented with the Excellence in Geophysical Education Award by the AGU, and in 2007 he was selected Virginia Outstanding Scientist. Dr. Hornberger is a member of the U.S. National Academy of Engineering, having been elected in 1996. He holds a B.S. in civil engineering and an M.S. in hydrology from Drexel University, and a Ph.D. in hydrology from Stanford University.

Kate Hadley Baker is retired after nearly 30 years of experience in the upstream industry and was recently the technology unit coordinator of Reservoir Performance for the BP Upstream Technology Group and the senior advisor of Subsurface for BP Group. From 1998 to 2001, Dr. Baker served as the director of Upstream Digital Business for BP Amoco, managing the consolidation of BP,

Amoco, and later, ARCO Upstream IT assets. In 1997, she served as manager of the Technical Services Group at Amoco Worldwide Exploration. There, she managed a group of 130-150 people providing hardcopy data to high-performance computing services. Other positions Dr. Baker held at Amoco include exploration manager of the Rockies Gas Business Unit, Amoco U.S. Operations/Energy Group, North America, and director of Geoscience Technology for Amoco Production Research. Dr. Baker joined Amoco to help form an upstream R&D organization that could do both research and technical services under one roof. From 1975 to 1994, she held many positions for Exxon Production Research Company and Exxon Company, USA, including research geologist, research supervisor, senior supervisory geologist, project leader, planning advisor, division operations geologist, well evaluation coordinator, and senior research supervisor. She has been a member of the Society of Petroleum Engineers (SPE) for 25 years and is a member of the Department of Energy's Council on Earth Sciences. Dr. Baker holds a B.S. in geology and a Ph.D. in geophysics from Massachusetts Institute of Technology.

Susan Brantley is distinguished professor of geosciences in the College of Earth and Mineral Sciences at Pennsylvania State University where she is also the director of the Earth and Environmental Systems Institute. She has been on the faculty at Penn State since 1986. Dr. Brantley's career as a geochemist focuses on the chemistry of natural waters both at the surface of the earth and deeper in the crust. Dr. Brantley and her research group investigate chemical, biological, and physical processes associated with the circulation of aqueous fluids in shallow hydrogeologic settings. She has published more than 160 refereed journal articles and 15 book chapters. Dr. Brantley is a fellow of the American Geophysical Union (AGU), the Geological Society of America (GSA), the Geochemical Society, the European Association of Geochemistry, and the International Association for GeoChemistry. She was president of the Geochemical Society from 2006 to 2008. Dr. Brantley was awarded the Arthur L. Day Medal from the GSA in 2011, the Presidential Award from the Soil Science Society of America in 2012, and an honorary doctorate from the Paul Sabatier University (Toulouse III, France) in 2012. Dr. Brantley was appointed to the U.S. Nuclear Waste Technical Review Board in 2012 by President Barack Obama. Also in 2012, she was elected to membership in the U.S. National Academy of Sciences. Dr. Brantley received her A.B. in chemistry (1980) and her M.A. and Ph.D. in geological and geophysical sciences in 1983 and 1987, respectively, all from Princeton University.

Michael E. Hohn was appointed state geologist and director of the West Virginia Geological and Economic Survey in Morgantown, West Virginia, in 2006. He has published more than 50 papers on energy resources and a book on geostatistics. His research interests include carbon sequestration, resource assessment, reservoir heterogeneity, and geostatistics. He was principal investigator for several projects funded by the U.S. Geological Survey, U.S. Department of Energy, the former Gas Research Institute, and the U.S. Forest Service. He served as treasurer, secretary general, and president for the International Association for Mathematical Geology (IAMG); secretary, vice president, and president for the Eastern Section of the American Association of Petroleum Geologists (AAPG); deputy editor of Natural Resources Research; editor-in-chief of Mathematical Geology; and treasurer for the Association of American State Geologists. His committee work has included the Energy Statistics Committee of the American Statistical Association, the Geologic Computing Committee of AAPG, and the Publication Committee of the IAMG. He is a fellow of the Geological Society of America. He holds a B.S. in geology from Binghamton University, and an M.S. and Ph.D. in geology from Indiana University.

Carl Kirby is professor of geology at Bucknell University where he also directs the Marcellus Shale Initiative. He has recently published abstracts, supervised student research, and is preparing articles on Marcellus Shale issues. Dr. Kirby has given numerous invited presentations and is collaborating with colleagues at the U.S. Geological Survey, University of Pittsburgh, and Binghamton University. He recently presented at an Environmental Protection Agency workshop on hydraulic fracturing. His research interests include alkalinity and acidity measurement and theory in mine drainage, municipal solid waste ash characterization and alteration, and mine drainage sediments for use as pigment. Dr. Kirby holds a B.S., an M.S., and a Ph.D. in geology from Virginia Polytechnic Institute and State University.

Appendix F

Acronyms and Abbreviations

BCF	billion cubic feet
BTEX	benzene, toluene, ethylbenzene, and xylene
BTU	British thermal unit
EIA	Energy Information Administration
STRONGER	State Review of Oil & Natural Gas Environmental Regulations
TCF	trillion cubic feet
USGS	U.S. Geological Survey
WINGS	Wildlife Incentives for Nongame and Game Species